T0340036

Wood, Trade, and Spanish Naval Power (*c*.1740–1795)

Brill's Series in the History of the Environment

Series Editor

Aleks Pluskowski, *University of Reading*

VOLUME 7

The titles published in this series are listed at *brill.com/bshe*

Wood, Trade, and Spanish Naval Power (*c.*1740–1795)

By

Rafał B. Reichert

With the Participation of

Karolina Juszczyk
Daniel Prusaczyk

BRILL

LEIDEN | BOSTON

Library of Congress Cataloging-in-Publication Data

Names: Reichert, Rafal, author.
Title: Wood, trade, and Spanish naval power (c.1740-1795) / by Rafał B.
Reichert ; with the participation of Karolina Juszczyk, Daniel
Prusaczyk.
Description: Leiden ; Boston : Brill, [2024] | Series: Brill's series in
the history of the environment, 1876–6595 ; volume 7 | Includes
bibliographical references and index. | Summary: "By focussing on timber
sourcing, this book sheds light on the exploitation of forests in
settings outside the Iberian Peninsula, including foreign states in the
southern Baltic region and the colonial territory of New Spain between
the c.1740-1795. Analysis of contracts, projects, and their
implementation by the Spanish crown in the 18th century allow for a
better understanding of the position of the Spanish monarchy's nearly
global efforts to sustain its naval commitments in the Atlantic World"–
Provided by publisher.
Identifiers: LCCN 2024010382 (print) | LCCN 2024010383 (ebook) | ISBN
9789004689633 (hardback) | ISBN 9789004689640 (ebook)
Subjects: LCSH: Lumber trade–Spain–History. | Spain–History, Naval.
Classification: LCC HD9765.S7 R453 2024 (print) | LCC HD9765.S7 (ebook) |
DDC 338.4/76740946–dc23/eng/20240422
LC record available at https://lccn.loc.gov/2024010382
LC ebook record available at https://lccn.loc.gov/2024010383

Typeface for the Latin, Greek, and Cyrillic scripts: "Brill". See and download: brill.com/brill-typeface.

ISSN 1876-6595
ISBN 978-90-04-68963-3 (hardback)
ISBN 978-90-04-68964-0 (e-book)
DOI 10.1016/9789004689640

Contents

Acknowledgements

Here, I would like to thank the individuals and the institutions who supported me during the writing and editing of this book.

First, I would like to thank my beloved Małgosia and Marysia, for their love, and constant support and for enduring my changing moods and hours of absence during the writing of this book. Also, to my parents Anna and Henryk, my sister Magda, my brother Andrzej, and other members of my family. As well as friends and colleagues from Mexico, Spain, and Poland for supporting my research and advising me during the writing of the manuscript. Especial thanks must go to Johanna von Grafenstein for her friendship from our doctoral days to today, when she often helped me on important wood-related issues, and Iván Valdez Bubnov, for the time spent on our maritime-naval discussions and projects, and for socializing at Coyoacán.

I would like to thank the National Science Centre for the funding granted to the project; the University of Warsaw, and especially the Dean's office (Robert Małecki and Anna Wojtyś) and the employees of the economic section of the Faculty of Modern Languages (Marta Zaręba and Izabela Poniatowska), for all their administrative help during the project and during the writing of this book.

I would like to thank my collaborators on the project Inez Beszterda, Łukasz Brylak, Emil Chróściak, Karolina Juszczyk, and Daniel Prusaczyk, whose professionalism helped the project to run smoothly, facilitated research and, finally, made the writing of this book possible.

Many thanks to María Isabel Rodríguez Ramos for the excellent editing and proofreading of the first draft of the text in Spanish and the professional English translation, editing, and proofreading made by David J. Govantes-Edwards.

Special thanks to my dear friends Hermano Ric, Amigocha Margarita, and Vecina Perla.

I thank the Centro de Estudios Superiores de México y Centroamérica of the Universidad de Ciencias y Artes de Chiapas, where I worked and where, for the first time, I had the opportunity to research the subject of the Spanish timber supply.

I would like to thank the archives of Mexico City (AGNM), Seville (AGI), Simancas (AGS), Madrid (Archivo Histórico de la Armada sede "Elcano"), and Warsaw (AGAD) for providing the maps and illustrations that enriched this book.

Finally, I would like to emphasize that this book is the main result of the project: "The Role of Wood Supplies from the Southern Baltic Region and the Viceroyalty of New Spain in the Development of the Spanish Seaborne Empire in the Eighteenth Century" (register no. 2017/25/B/HS3/01017), which was funded by the National Science Centre, Poland.

Figures, Maps and Tables

Figures

Maps

Tables

FIGURE 1   Image of the cedar tree from the map of San Miguel Chiquistepeque y San Francisco Puctla, New Spain (1757) (Archivo General de la Nación de México, Mapas, Planos e Ilustraciones (280), MAPILU/210100/2322 San Miguel Chiquistepeque y San Francisco Puctla; Minas de Sacualpa. Edo. de Méx., 2231)

Introduction

The dynastic change in Spain in 1700 brought about a series of political, economic, military, and cultural changes. The new Bourbon dynasty, represented by King Philip V (1700–1746), set about restoring and strengthening the Spanish Empire against its main rival, Great Britain. Key to this ambitious project was the modernisation of the navy, which in the words of John Lynch: "had not even a good place to boil a cauldron of pitch".[1] This lamentable state of the Spanish navy was the consequence of the neglect of the shipbuilding industry during the reign of the last Habsburg, Charles II (1665–1700), and also the result of the War of the Spanish Succession (1701–1713) during which the last of the king's ships were immobilised. For this reason, the mission of reviving Spanish naval power began as early as 1714. The creation of the Ministry of the Royal Navy and the West Indies, and later of the intendancies, maritime departments, and royal shipyards in Guarnizo, El Ferrol, Cádiz-La Carraca, Cartagena de Levante, and Havana laid the foundations for the development of modern shipbuilding in Spain. Among the various figures who contributed to this process, the most outstanding are José Patiño, intendant and later minister of the Royal Navy, and Antonio de Gaztañeta, a sailor and shipbuilder who drew up the regulations for the king's shipbuilding system in the "Spanish style".[2] These efforts by the king, his naval officers, and the royal treasury led to the revival of naval power, and by the eve of the War of Jenkins' Ear (1739–1748) the Spanish fleet comprised 29 ships-of-the-line, 11 frigates, six *paquebotes*, four bombards, two galleons, two *azogues*, two *galeotas*, two sloops, and one *pingüe*.[3]

This policy of naval reinforcement continued between 1743 and 1795 under the Marquis of Ensenada (1743–1754), Julián de Arriaga (1754–1776), Pedro González de Castejón (1776–1783), and Antonio Valdés (1783–1795), José Patiño's successors in the Ministry of the Royal Navy. Owing to this successful strategy, by the 1790s the Spanish Navy was the second most powerful navy in Europe, after the British Royal Navy. Several factors contributed to this: financial stability; the skilful management of the intendants and officers of the *Marina Real* and the maritime departments; the implementation of technological

---

1 John Lynch, *Historia de España. Edad moderna: crisis y recuperación 1598–1808*, vol. V (Barcelona: Crítica, 2005), 389.
2 Iván, Valdez-Bubnov, *Poder naval y modernización del Estado: política de construcción naval española (siglos XVI–XVIII)* (México: UNAM, 2011), 140–147.
3 Ana Crespo Solana, *La Casa de Contratación y la Intendencia General de la Marina en Cádiz (1717–1730)* (Cádiz: Universidad de Cádiz, 1996), 45.

© KONINKLIJKE BRILL BV, LEIDEN, 2024 | DOI:10.1163/9789004689640_002

innovations; and, finally, a functional raw material supply system. Jan Glete,[4] followed by Iván Valdez Bubnov[5] and Rafael Torres Sánchez,[6] pointed out that this policy was grounded in the broader military strategy pursued by the Spanish and French crowns, whereby the Bourbons' combined fleets[7] could muster as many ships-of-the-line and frigates as the British navy.[8] This status was reached after the end of the War of Jenkins' Ear (1748), and in the period between 1770 and 1789, the naval forces of Spain and France outstripped the Royal Navy in terms of warships. Thus, in the second half of the 18th century, the Bourbon navies became the mainstay of the colonial order, contributing to the naval balance against the powerful British navy. However, it is also clear that, despite having a similar or smaller number of ships, supremacy in combat was on the British side, because their crews and officers were better trained. This is why the Franco-Hispanic fleets meet with more defeats than victories in the second half of the 18th century.

British policies to strengthen the Royal Navy during this period triggered the intensification of naval production in Spain and was also reflected in the introduction of technological innovations in shipbuilding. In the second half of the 18th century, the shipbuilding industry produced vessels based on three systems: the "English style", introduced by Jorge Juan and the British constructors hired by the Bourbon Crown (1750–1765); the "French style", implemented in the royal shipyards by the shipbuilder François Gautier (1765–1782); and the "mixed or perfected style", led by the naval constructors José Romero de Landa and Julián Martín de Retamosa, who in their designs combined features of the earlier systems, for example using rigging in the English style to increase the speed of ships.[9]

---

4   Jan Glete, *Navies and Nations: Warships, Navies and State Building in Europe and America, 1500–1860* (Stockholm: Almqvist & Wiksell International, 1993), 553–579.

5   Valdez-Bubnov, *Poder naval y modernización del Estado*, 329–343.

6   Rafael Torres Sánchez, *Historia de un triunfo. La Armada española en el siglo XVIII* (Madrid: Desperta Ferro, 2021), 126.

7   Close collaboration between the Spanish and French Bourbons was the outcome of the Family Pact, which was signed on three occasions: in 1733, 1743, and 1761. It should be noted that, in 1779, the Third Family Pact was renewed by the Treaty of Aranjuez and remained in force until the outbreak of the French Revolution in 1789.

8   During the 18th century, the Royal Navy maintained between 100 and 140 major ships (ships-of-the-line and frigates). Torres Sánchez, *Historia de un triunfo*, 126.

9   José Patricio Merino Navarro, *La Armada Española en el siglo XVIII* (Madrid: Fundación Universitaria Española, 1981), 130–136; José M. de Juan-García Aguado, *José Romero Fernández de Landa: un ingeniero de marina en el siglo XVIII* (La Coruna: Universidade da Coruña, 1998), 184–189.

The analysis of Bourbon Spain's naval policy in the second half of the 18th century clearly shows that the most valuable raw material for the ambitious plans to modernise the *Marina Real* was wood. For this reason, the main subject of this book is timber sourcing, with special attention to the exploitation of forests outside the Iberian Peninsula, including foreign states in the southern Baltic region and in the Spanish colonies, for instance the viceroyalty of New Spain. This book describes the different ways in which wood supply operated, depending on local geographical factors, political situation, available personnel and funds, and the broader context of naval affairs involving the Spanish state, its allies, and its enemies. That is why the research presented in this book starts with two important questions: (1) how did Spain keep up with the increasing demand for timber as its naval power surged in the second half of the 18th century? and (2) what were the strategies deployed by the Bourbon monarchy to exploit and capture forest resources outside the Iberian Peninsula?

In order to answer these two questions, it is necessary to revise the Spanish Crown maritime policies and the development of its naval power, which was the key factor in the protection of, and trade with, its overseas colonies in America. Before, however, it is worth recalling that, in early modern Europe, ships became the main medium of colonial expansion, and their production essential for the economic and military stability of crowns, states, and nations. European demand for wood gradually began to grow from the 1640s, and became acute during the 18th century, mostly as a result of naval competition between England/Great Britain, France, Spain, Holland, and Portugal. The resulting overexploitation of woodland led to the deforestation of many regions in these countries,[10] forcing naval powers to reconnoitre other

---

10   See for instance Manuel Corbera Millán, "El impacto de las ferrerías en los espacios forestales (Cantabria, 1750–1860)", *Ería* 45 (1998), 89–102; Gaspar de Aranda y Antón, *El camino del hacha: la selvicultura, industria y sociedad: visión histórica* (Madrid: Ministerio de Medio Ambiente, 1999); John Langton & Jones, Graham (eds), *Forests and Chases of England and Wales c.1500–1850: Towards a Survey and Analysis* (Oxford: St John's College, 2005); Hamish Graham, "For the Needs of the Royal Navy: State Interventions in the Communal Woodlands of the Landes during the Eighteenth Century", *Journal of the Western Society for French History* 35 (2007), 135–148 and "Fleurs-de-lis in the Forest: 'Absolute' Monarchy and Attempts at Resource Management in Eighteenth-Century France", *French History* 23, no. 3 (2009), 311–335; Keiko Matteson, *Forests in Revolutionary France: Conservation, Community, and Conflict, 1669–1848* (Cambridge: Cambridge University Press, 2015); Koldo Trapaga Monchet, "Guerra y deforestación en el reino de Portugal (siglos XVI–XVII)", *Tiempos Modernos* 39, no. 2 (2019), 396–425; Koldo Trápaga Monchet and Félix Labrador Arroyo, "Políticas forestales y deforestación en Portugal, 1580–1640: realidad o mito?", *Ler História* 75 (2019), 133–156; Koldo Trapaga Monchet, Álvaro Aragón-Ruano,

European markets for timber and other sources, from the hinterlands of the Baltic, Adriatic, and Black Sea regions to America and Asia.[11]

No raw material – apart from precious metal – was as highly coveted by European naval powers as timber. In the Early Modern Age, this forest resource became the main tool for expansionist and colonialist policies, allowing European naval powers to develop their economy, trade, and military power. For this reason, greater emphasis needs to be put on wood as a strategic resource that led to technological progress in shipbuilding and the modernisation of the navies of England/Great Britain, Spain, France, Holland, and, to a lesser extent, Portugal, Sweden, and Denmark. This is not only from the political – military perspective but, as the Spanish case illustrates, also in terms of knowledge about forest resources in Iberian Peninsula and the American colonies. Thanks to the explorations and surveys undertaken by the Spanish army and naval officers,[12] the rational harvesting of wood – to ensure the navy's ability to self-supply – became a target in itself, and the crown implemented more

Cristina Joanaz de Melo, *Roots of Sustainability in the Iberian Empires. Shipbuilding and Forestry, 14th–19th Centuries* (New York: Routledge, 2023).

11   See for instance D. R. Radell, and J. Parsons, "El Realejo: A Forgotten Colonial Port and Shipbuilding Center in Nicaragua", *Hispanic American Historical Review* 51 (1971), 295–312; Warren Dean, *With Broadax and Firebrand: The Destruction of the Brazilian Atlantic Forest* (Berkeley: University of California Press, 1995); Peter Boomgaard, "The VOC Trade in Forest Products in the Seventeenth Century", in Richard Grove, Vinita Damodaran, and Satpal Sangwan (eds), *Nature and the Orient: The Environmental History of South and Southeast Asia* (Delhi: Oxford University Press, 1998); Reinaldo Funes Monzote and Reinaldo Funes Monzote, "Los conflictos por el acceso a la madera en La Habana: hacendados vs Marina", in José Piqueras (ed.), *Diez nuevas miradas de historia de Cuba* (Castellón: Publicacions de la Universitat Jaume l, 1998), 67–90 and *From Rainforest to Cane Field in Cuba: An Environmental History since 1492* (Chapel Hill: University of North Carolina Press, 2008); Shawn W. Miller, *Fruitless Trees: Portuguese Conservation and Brazil's Colonial Timber* (Stanford, CA: Stanford University Press, 2000); Greg Bankoff, "The Tree as the Enemy of Man: Changing Attitudes to the Forests of the Philippines, 1565–1898", *Philippine Studies* 52, no. 3 (2004), 320–344; Germán Luis Andrade Muñoz, *Un mar de intereses, la producción de pertrechos navales en Nueva España, siglo XVIII* (México: Instituto Mora, 2006); Miguel Jordán Reyes, "La deforestación de la Isla de Cuba durante la dominación española: (1492–1898)" (PhD dissertation, Universidad Politécnica de Madrid, Madrid, 2006); Greg Bankoff and Peter Boomgaard (eds), *A History of Natural Resources in Asia: The Wealth of Nature* (Basingstoke: Palgrave Macmillan, 2007); Iván Valdez-Bubnov, "La construcción naval española en el Pacífico sur: explotación laboral, recursos madereros y transferencia industrial entre Nueva España, Filipinas, India y Camboya (siglos XVI y XVII)", *Studia Historica-Historia moderna* 43, no. 1 (2021), 71–102.

12   John Robert McNeill, "Woods and Warfare in World History", *Environmental History* 9, no. 3 (2004), 388–410; Valdez-Bubnov, *Poder naval y modernización del Estado*; Rafael Torres Sánchez, *Military Entrepreneurs and the Spanish Contractor State in the Eighteenth Century* (Oxford: Oxford University, 2016).

stringent monitoring systems to maintain the sourcing of this strategic raw material under strict control of the state.[13]

From the 16th century onwards, the colonial expansion of the European crowns in Africa, America, and Asia allowed the exploration of forests on these continents. This event opened up the opportunity to exploit new tree species other than those traditionally used for centuries in European shipbuilding, which were more suitable for sailing in warmer seas. The trees commonly used in the European shipbuilding industry were pine, oak, beech, spruce, black poplar, and elm. Depending on their physical characteristics (weight, strength, durability, density, elasticity), these European species were used to manufacture different ship parts. For example, pines were mainly used for masting and, to a lesser extent, for planking; beech for oars and rudders; black poplar and elm for planks, ribbons, and turn pieces; oak and melis pine for beams; and pedunculate oak for the structural parts of ships, such as keels, frames, elbows, and rods.[14]

The knowledge about the properties of tropical wood allowed the progress of European shipbuilding, which began to increasingly use American and Asian species, such as mahogany, sabicú, cedar, cypress, and also teak, ebony, guijo, betis, banaba: woods that performed better in warm waters than European timber, as well as being more resistant to naval shipworm or *turu*.[15] For this reason, their service life was longer than those of vessels built with European

---

13 See for instance Luis Urteaga, *La tierra esquilmada. Las ideas sobre la conservación de la naturaleza en la cultura española del siglo XVIII* (Madrid: Serbal/CSIC, 1987); Gaspar Aranda y Antón, *Los bosques flotantes. Historia de un roble del siglo XVIII* (Madrid: ICONA, 1990); Ofelia Rey Castelao, *Montes y política forestal en la Galicia del Antiguo Régimen* (Santiago de Compostela: Universidad de Santiago de Compostela, 1995); Pilar Pezzi Cristóbal, "Proteger para producir. La política forestal de los Borbones españoles", *Baetica. Estudios de Arte, Geografía e Historia* 23 (2001), 583–595; Álvaro Aragón Ruano, *El bosque guipuzcoano en la Edad Moderna: aprovechamiento, ordenamiento legal y conflictividad* (Donostia: Sociedad de Ciencias Aranzadi, 2001) and "Siete siglos de sostenibilidad forestal en Guipúzcoa (siglos XIII–XIX)", *Manuscrits. Revista d'Història Moderna* 42 (2020), 65–88 and "Soberanía y defensa de la riqueza forestal en la frontera vasconavarra con Francia durante el siglo XVIII", *Memoria y Civilización* 25 (2022), 423–450; John T. Wing, *Roots of Empire: Forests and State Power in Early Modern Spain, c.1500–1750* (Leiden: Brill, 2015); Alfredo José Martínez González, *Las Superintendencias de Montes y Plantíos (1574–1748). Derecho y política forestal para las armadas en la Edad Moderna* (Valencia: Tirant Lo Blanch, 2015); Vicente Ruiz García, "La Provincia Marítima de Segura (1733–1836). Poder Naval, Explotación Forestal y Resistencia en la España del Antiguo Régimen" (PhD dissertation, Universidad de Murcia, Murcia, 2018).

14 Gaspar de Aranda y Antón, *La carpintería y la industria naval en el siglo XVIII* (Madrid: Instituto de Historia y Cultura naval, 1999), 23–24.

15 *Teredo navalis*, an elongated, worm-shaped mollusc that bored through ship hulls.

wood. Once this critical issue was well understood, the navies of Spain, France, England/Britain, Portugal, and Holland began taking advantage of their colonies to exploit the durable tropical woods for the construction and repair of their ships overseas and in their European shipyard.[16]

Spain became aware of the importance of timber early, after the *Carrera de Indias*, which began systematically in 1543, triggered a constant demand for merchant ships to channel trade with the colonies and armed galleons to protect the vital routes along which precious metals and other commodities circulated.[17] In 1562, Philip II and the councils of the Indies and State took measures to begin naval production on a mass scale to create the armadas needed to protect the vast Spanish Empire. This required the reorganisation of forest management by royal officials and massive tree-planting campaigns. In 1574, Philip II appointed Cristóbal de Barros y Peralta first Superintendent of Hills and Forests. The promotion and protection of woodland in northern Spain can be regarded as the first expression of sustainable forestry policies, which was not grounded on conservationist ideas, but on a strategy to guarantee a constant source of raw materials to build and repair ships.[18] Interestingly, the same notions guided Spanish public policies until the 1830s, when orders to preserve and manage forests (e.g. *Ordenanzas Generales de Montes*, 1833) were published on the initiative of Minister Javier de Burgos. Importantly, his orders imposed strict restrictions on logging as an urgent measure to stop soil erosion and flooding, which were exacerbated by the lack of forested areas. For the first time in Spanish history, these ordinances reflected a change in the reasoning behind forestry policies: from preservation with a view on naval construction to the protection of woodland for conservationist purposes.[19]

Back to the Early Modern Age, it is worth recalling that, between 1574 and the early 19th century, Spanish monarchs and officials triggered several policies to impose control over the exploitation of forest resources in Spain and the colonies. Examples of these official regulations, which focused on the

---

16   Eugenio Plá y Rave, *Tratado de maderas de construcción civil y naval* (Madrid: Imprenta, Estereotipia y Galvanoplastia de Aribau, 1880), 124; Greg Bankoff and Peter Boomgaard (eds), *A History of Natural Resources in Asia: The Wealth of Nature* (Basingstoke: Palgrave Macmillan, 2007), 13–16; Reinaldo Funes Monzote, *From Rainforest to Cane Field in Cuba: An Environmental History since 1492* (Chapel Hill: University of North Carolina Press, 2008), 7–38.

17   Clarence H. Haring, *Trade and Navigation between Spain and the Indies in the Time of the Hapsburgs* (Cambridge, MA: Harvard University Press, 2014 [1st edn 1918]), 251.

18   Alfredo José Martínez González, *Las Superintendencias de Montes y Plantíos (1574–1748). Derecho y política forestal para las armadas en la Edad Moderna* (Valencia: Tirant Lo Blanch, 2015), 46–47.

19   Ignacio García Pereda, Inés González Doncel, and Luis Gil Sánchez, "La primera Dirección General de Montes (1833–1842)", *Quaderns d'Història de l'Enginyeria* 13 (2012), 215–230.

use of common woodland – exploitation regimes, fines and punishments for non-compliance, and the impact of agriculture and stock-breeding – can be found in the Catalonian *Ordinacions Forestals*, written in 1627 by Miguel de los Santos de San Pedro. These rules tried to impose severe limitations on logging, while reinforcing the presence of the navy commissars tasked with selecting the most suitable trees for naval construction.[20]

Another example is Toribio Pérez de Bustamante's *Instrucción forestal* (1656), with which Philip IV tried to centralise forest management not only in Spain but in his whole empire. The document established a new basic rule for the preservation of woodland: the prohibition to fell trees within two leagues of the coastline and navigable rivers, which were put under royal protection. The *Instrucción* also ordered local authorities to disseminate the document's contents in the areas in which these forests were. Interestingly, churches were singled out as ideal points to disseminate the order. The *Instrucción* also included practical information, concerning felling periods, tree planting, and punishments for those that violated the ban on logging in royal, communal, and private woodland.[21]

It is interesting that the Spanish Habsburgs also included their American colonies in this legislation. Unsurprisingly, the first ordinance to protect American woodland was passed in Cuba, by order of Philip II in 1559. This regulation prohibited any logging, two leagues inland on the banks of the Chorrera River and also within ten leagues leeward and windward of Havana, without the governor's permission. Anyone caught with an axe or a machete in the forests was sentenced to forced labour at the fortification works. Although this first order does not mention the navy's needs, the following do, because throughout the 17th century the Spanish Crown prioritised shipbuilding in Cuba, notably in Havana. Especially important among all the royal orders concerning the navy's wood resources was the one issued by Philip III on 26 March 1607, ordering the Cuban governor, Pedro de Valdés, to send 50 pieces and 100 boards of mahogany to the Casa de Contratación, Seville, for its qualities as shipbuilding material to be assessed. Also important was the order issued on 2 March 1620, authorising anyone arriving in Havana with the intention of building ships to fell trees anywhere in Cuba, and that issued on 2 March 1623, with which Philip IV addressed the petition of Cuba's governor, Francisco de Venegas, to limit this permit to the constructors of *naos* only, because *vecinos* of Havana, especially

---

20   Vicente Casals Costa, "Conocimiento científico, innovación técnica y fomento de los montes durante el siglo XVIII", in Manuel Silva Suárez (ed.), *El Siglo de las luces: de la industria al ámbito agroforestal* (Zaragoza: Institución Fernando el Católico, Universidad de Zaragoza-Real Academia de Ingeniería, 2005), 457–459.
21   Wing, *Roots of Empire*, 141–145.

ranchers, were expanding their land at the expense of woodland that was valuable for naval construction.[22]

During the rule of the Spanish Bourbons, one of the most important decisions was to publish the *Ordenanza de 1748*, which imposed norms for the exploitation of woodland in the Iberian Peninsula under the supervision of superintendents from the navy departments of El Ferrol, Cádiz-La Carraca, and Cartagena. The main promoter of this document was the Marquis of Ensenada, Minister of the Royal Navy and the Indies, who granted navy officers sweeping powers to manage public and private woodland in the Iberian Peninsula, prioritising the Navy's needs and shipbuilding in the royal shipyards. They were bestowed with the authority to decide the fate of all woodland, and, in their surveys, they earmarked the best trees for the Spanish Navy, forcing private landowners to use inferior quality timber for their own needs. Intendants also had precedence to buy timber, sometimes paying below-market prices. Similarly, Navy officers had the authority to send supervisors to oversee planting operations.[23] These policies, however, did not yield the expected results, sparking conflict between state officials and forest owners,[24] and ultimately forcing the Ministry of the Royal Navy and Indies to seek other sources of timber outside the Iberian Peninsula through the *asiento* system.[25]

The analysis of these and other official documents,[26] issued during the Early Modern Age, clearly illustrate that policies to protect woodland in Spain were

22   Jordán Reyes, "La deforestación de la Isla de Cuba durante la dominación española", 46–50.

23   Alfredo José Martínez González, "La elaboración de la Ordenanza de Montes de Marina, de 31 de enero de 1748, base de la política oceánica de la monarquía española durante el siglo XVIII", *Anuario de Estudios Americanos* 71, no. 2 (2014), 571–602; Wing, *Roots of Empire*, 206–215.

24   For an example see Álvaro Aragón Ruano, "Un choque de jurisdicciones. Fueros y política forestal en el Pirineo occidental durante el siglo XVIII", *Obradoiro de Historia Moderna* 28 (2019), 135–162.

25   Rafal Reichert, "¿Cómo España trató de recuperar su poderío naval? Un acercamiento a las estrategias de la Marina Real sobre los suministros de materias primas forestales provenientes del Báltico y Nueva España (1754–1795)", *Espacio, tiempo y forma. Serie IV, Historia moderna* 32 (2019), 73–102; Rafael Torres Sánchez, "Los negocios con la armada. Suministros militares y política mercantilista en el siglo XVIII", in Iván Valdez-Bubnov, Sergio Solbes Ferri and Pepijn Brandon (eds) *Redes empresariales y administración estatal: la provisión de materiales estratégicos en el mundo hispánico durante el largo siglo XVIII* (México: UNAM, 2020), 49–76; Rafael Torres Sánchez, "Mercantilist Ideology versus Administrative Pragmatism: The Supply of Shipbuilding Timber in Eighteenth-Century Spain", *War & Society* 40, no. 1 (2021), 9–24.

26   It is important to emphasise that similar regulations were imposed in other European countries. See for instance Koldo Trápaga Monchet, "El estudio de los bosques reales de

guided by the wish to ensure the supply of timber for the king's ships. This policy of sustainable use of forest resources pursued by Bourbon Spain was in line with mercantilist ideas put forward in the first half of the 18th century by two Spanish politicians inspired by the philosophy of the Enlightenment, Jerónimo de Uztáriz and José del Campillo y Cossío, who highlighted the huge potential of American forests as the main source of timber for the Spanish royal shipyards. Uztáriz saw great advantage in the durability and resistance of American timber, which he expressed as follows:

> In the islands and mainland of America, where his majesty has many exquisite kinds of wood and an abundance of pitch and tar for the construction of ships ... with the considerable benefit that if the [ships] made in Europe resist for 12 to 15 years, [those in America] are preserved for more than 30 years since they are made there with the cedar, harder oak, and other woods of superior strength and resistance.[27]

Campillo y Cossío also demonstrated the advantages of American wood, saying that:

> Campeche wood, cedar, mahogany and other beautiful woods, masts for ships, planks, pitch ... that now come to us from the Baltic, we will have from our Indies; and also the furniture, tools, instruments for work ... we can take them [from Spain] there and sell them cheaper [in America].[28]

Portugal a través de la legislación forestal en las dinastías Avis, Habsburgo y Braganza (c.1435–1650)". More directly in relation to the chronological framework of this book, the second half of the 18th century, several royal orders in Prussia (1750 and 1777) and the Polish-Lithuanian Commonwealth (1775 and 1778) already present significant differences with the Spanish regulation of 1748, because the Prussian and Polish orders aimed to protect woodland from deforestation, rather than ensuring raw materials for state policies.

27 Jerónimo de Uztáriz, *Theorica y practica de comercio, y de marina: en diferentes discursos y calificados exemplares, que con específicas providencias, se procuran adaptar á la Monarquia española* (Madrid: Imprenta de Antonio Sanz, 1757), 216:"son grandes las ventajas que en las islas y tierra firme de la América tiene su majestad de muchas y exquisitas maderas y abundancia de brea y alquitrán para la construcción de bajeles... con el considerable beneficio de que si los [buques] fabricados en Europa duran de 12 a 15 años, [estos de América] se conservan más de 30, ya que se hacen allá con el cedro, roble más duro y otras maderas de superior firmeza y resistencia."

28 José del Campillo y Cossío, *Nuevo sistema de gobierno económico para la América: con los males y daños que le causa el que hoy tiene, de los que participa copiosamente España; y remedios universales para que la primera tenga considerables ventajas, y la segunda mayores intereses* (Madrid: Imprenta Benito Cano, 1789), 158. "El palo de Campeche, cedro,

His idea clearly shows that Spain should replace the timber provision from the Baltic and also from other European regions because the use of American timbers would offer a double benefit. First, the money invested in the purchase of timber production in the colonies would remain in Spanish commercial circuits. Second, timber extraction would be supervised by the royal and colonial authorities and would guarantee quality control of the wooden parts produced for ships. Uztáriz and Campillo y Cossío's ideological approach was materialised during logging operations in Oaxaca and Cuba, which were carried out under the patronage of the Crown, and the concession of *asientos* to merchants and influential Creole residents from New Spain, Louisiana, and Cuba during the second half of the 18th century (see Chapter 4).

During this period, especially after the Treaty of Paris (1763), which brought an end to the Seven Years' War, in which Spain was defeated by the British, strenuous efforts were made to reinforce the Spanish Navy, which after the time of Ensenada had entered a period of stagnation. The first major change was technological, with the substitution of the "French style" of shipbuilding for the "British system", which had been followed in the 1750s and the early 1760s. In 1772, Francisco Gautier issued a report entitled *Observaciones sobre el estado de los montes de España, nota del consumo de la madera de construcción, que, en cada año, se considera necesaria en los departamentos de Ferrol, Cartagena y Cádiz; y proyecto para aprovisionar estos arsenales de maderas de América,*[29] in which he described the state of Spanish woodlands and made audacious proposals to use American timber in naval construction in the Iberian Peninsula. Gautier emphasised that "Cádiz uses American wood which is brought at great cost but with little profit, not because it lacks in quality, but because of the carelessness with which it is dispatched [to Spain], where it arrives badly cut, badly arranged, and poorly sorted".[30] These words somewhat concealed his

---

caoba y otras maderas hermosas, mástiles para navíos, tablazón, brea, pez... que ahora nos vienen del Báltico, los tendremos de nuestras Indias; y asimismo los muebles, herramientas, instrumentos para labor... los podremos llevar [de España] allá y venderlos barato [en América]".

29   "Observations on the state of the mountains of Spain, note of the consumption of construction wood, which, in each year, is considered necessary in the departments of Ferrol, Cartagena, and Cádiz; and project to supply these arsenals with wood from America".

30   [Cádiz utiliza la madera de América que viene a mucha costa y de la cual se saca muy poco provecho, no por su calidad, que es de preferir a cuanto roble hay en España, sino por el ningún cuidado de los que la envían a [la metrópoli], donde llega muy mal delineada, mal configurada y nunca surtida]. AGS, SMA, Arsenales, leg. 349. "Observaciones sobre el estado de los montes de España, nota del consumo de la madera de construcción, que, en cada año, se considera necesaria en los departamentos de Ferrol, Cartagena y Cádiz; y proyecto para aprovisionar estos arsenales de maderas de América".

real intentions, which were to base the Navy's timber supplies mostly on the exploitation of colonial forests; he points out that there was already a flow of *tozas*[31] between Havana and other naval departments, but that this was insufficient and was limited to specific ship parts of sabicú, mahogany, and cedar.

In any case, Gautier's project began initiatives by the Navy and viceregal authorities in the colonies to exploit American forests. In the 1770s and 1780s, various woodland surveys were launched by royal officials to inspect potential timber sources, establishing species and volumes, from the Greater Caribbean and Louisiana to Veracruz, Oaxaca, Yucatán, Darién, the Magdalena River, Cumaná, and the Orinoco. Not all of these crystallised in felling *asientos*, but others met with greater success. As such, Louisiana and Chimalapas became sources of pine masting used in Havana; Coatzacoalcos, Tlacotalpan, Alvarado, Laguna de Términos, Cartagena de Indias, the Magdalena River, and Cumaná supplied shaped pieces and boards in cedar and mahogany, most of which were shipped to the Iberian Peninsula. Gautier's idea to tap into the enormous resources of American forests made sense, as the shipbuilder wished to keep Spanish woodland as a strategic reserve for naval construction. However, the project was hampered by difficulties in hauling shaped pieces timber to Spain and could only become a complement to the supplies obtained in the Iberian Peninsula and other European regions.

In the second half of the 18th century, the main supply routes for the Spanish Navy linked with the Baltic and the Northern seas, where timber was obtained from private merchants,[32] mostly foreigners and representatives of trade houses from the Netherlands, England, France, and Scandinavia, which frequently had branches in Spain. These businesspeople were connected with Spanish traders and their political allies in Madrid and the main Spanish harbours (Bilbao, San Sebastián, Santander, Cádiz, Seville, Malaga, Cartagena, and Barcelona). The reason behind this model was the shortcomings of the Spanish merchant navy, which from the 17th century was gradually replaced by cheaper foreign freight options – depending on geopolitical conditions, English/British, Dutch, French, and, to a lesser extent, Scandinavians and Hamburgers. In the Mediterranean, the second major source of wood (Romanian, Balkan, and

---

31   Eng. blocks.
32   Germán Jiménez-Montes, *A Dissimulated Trade: Northern European Timber Merchants in Seville (1574–1598)* (Leiden: Brill, 2022). It should also be noted that several archives – General de Simancas, General de la Marina-Álvaro Bazán, Histórico Nacional, Provincial de Cádiz, and Naval de Cartagena – possess vast records about the purchase of timber by the Navy in Romania, Dalmatia, Albania, Italy, and the Papal States, in the Mediterranean; and the southern Baltic harbours, seen *in extenso* in this monograph, and other harbours in Russia, Sweden, Denmark, Norway, Hamburg, and Amsterdam, in the north.

Italian), most contracts were signed with Italian merchants,[33] who virtually monopolised timber supplies for the Spanish navy in the 18th century in the region, as illustrated by the contracts signed with *asentistas* Joseph Marcerano, Baltazar Castellini, and Carlos María Marraci.[34]

However, the most important source of wood during this period lay in the north, especially the Baltic, where Spain had to compete with other countries that traditionally sourced their timber there, that is, the British, Dutch, and French. Interestingly, with the help of a vast network of Spanish and foreign merchants and the diplomatic support of consuls, plenipotentiary ministers, and ambassadors, Bourbon Spain managed to gain, and keep, a foothold in the Baltic timber market in the period spanning the 1760s to the 1790s. This activity peaked in the run-up to the American War of Independence (1775–1783), which Spain joined in 1779, and in the second half of the 1780s. Especially interesting are the attempts to nationalise wood *asientos* instead of working with foreign merchants. The contracts awarded to Miguel Soto and his representative Felipe Chone in the 1770s were the first step, followed in 1782 by the creation of the Banco de San Carlos to centralise naval supplies and finances. The Soto-Chone *asiento* yielded some positive results, but the monopolistic position awarded to the national bank was a fiasco, and traditional policies to

---

33   See for instance Rafael Torres Sánchez, "La colonia genovesa en Cartagena durante la Edad Moderna", in Rafaela Belvederi (ed.), *Rapporti Genova-Mediterraneo-Atlantico nell'Età Moderna* (Genoa: Università di Genova, 1990), 553–581; Manuel Bustos Rodríguez, *Los comerciantes de la Carrera de Indias en el Cádiz del siglo XVIII* (Cádiz: Universidad de Cádiz, 1995); Ana Crespo Solana, *El comercio marítimo entre Cádiz y Ámsterdam, 1713–1778* (Madrid: Banco de España, 2000) and *Entre Cádiz y los Países Bajos: una comunidad mercantil en la ciudad de la Ilustración* (Cádiz: Fundación Municipal de Cultura, Cátedra Adolfo de Castro, 2001) and *Comunidades transnacionales. Colonias de mercaderes extranjeros en el Mundo Atlántico (1500–1830)* (Madrid: Doce Calles, 2010); Manuel Díaz Ordóñez, "El riesgo de contratar con el enemigo. Suministros ingleses para la Armada Real española en el siglo XVIII," *Revista de Historia naval* 21, no. 80 (2003), 65–74.

34   Joseph Marcerano was awarded an *asiento* to supply the department of Cartagena with Romanian wood for six years, beginning on 1 January 1761. Baltazar Castellini and Carlos María Marraci worked with the naval department of Cartagena and, in the case of Marraci, also with Felipe Chone, who purchased Baltic timber for El Ferrol, Cádiz-La Carraca, and Cartagena. Archivo General de la Marina-Álvaro de Bazán (Hereafter, AGMAB), Arsenales, legajo 3762; Archivo Naval de Cartagena (hereafter, ANC), Junta Económica del Departamento (hereafter, JED), libros de acuerdos vol. 2729, tomo 2 (1774) and Archivo General de Simancas (hereafter, AGS), Secretaria de Marina (Hereafter, SMA), Arsenales leg. 635 "Don Carlos María Marraci y Compañía, vecinos y del Comercio de esta Corte (1779)". It should be mentioned that timber trade in the Mediterranean is not presented in this work. However, as the above examples indicate, a considerable number of contracts and official documents are found in the Spanish archives that can be used to study this issue in the future.

find foreign merchants with solid Baltic networks, such as the Swedish Gahn, had to be resumed.[35]

It must be emphasised that the policies implemented by the Spanish Navy in the second half of the 18th century clearly show that Mahan's thesis[36] is – at least for this period – incorrect, because, as also shown by other recent research, Bourbon Spain had enough human capital, skilled royal officials, cash, natural resources in Spain and America, commercial networks, and diplomatic leverage to manage wood supplies in different northern and southern European markets, which led the Spanish armed forces to their greatest success in the 18th century, victory over Great Britain in the American War of Independence. This would have been impossible without the modernisation and development of the Navy, which in the 1780s became the second most powerful in the world, after the British[37].

As noted, the main aim of this book – which presents the results of research undertaken over ten years – was to contribute to our understanding of timber supplies for the Spanish Navy, especially from the southern Baltic and the viceroyalty of New Spain, for which little previous research has been undertaken. Although things have improved in recent years, significant gaps in our knowledge concerning these two regions still exist.[38] As such, a comprehensive

---

35   María Guadalupe Carrasco González, "Cádiz y el Báltico. Casas comerciales suecas en Cádiz (1780–1800)", in Alberto Ramos Santana *Comercio y Navegación entre España y Suecia (siglos X–XX)* (Cádiz: Universidad de Cádiz, 2000), 330–331 and Torres Sánchez, *Military Entrepreneurs*, 189–229.

36   Alfred T. Mahan, *The Influence of Sea Power upon History, 1660–1783* (Cambridge: Cambridge University Press, 2010 [1st edn 1889]).

37   José Patricio Merino Navarro, *La Armada Española en el siglo XVIII* (Madrid: Fundación Universitaria Española, 1981); Fernando Serrano Mangas, *Función y evolución del galeón en la Carrera de Indias* (Madrid: Mapfre, 1992); Alberto Ramos Santana *Comercio y Navegación entre España y Suecia (siglos X–XX)* (Cádiz: Universidad de Cádiz, 2000); Germán Luis Andrade Muñoz, *Un mar de intereses, la producción de pertrechos navales en Nueva España, siglo XVIII* (México: Instituto Mora, 2006); Iván, Valdez-Bubnov, *Poder naval y modernización del Estado: política de construcción naval española (siglos XVI–XVIII)* (México: UNAM, 2011).

38   Rafal Reichert, "El comercio directo de maderas para la construcción naval española y de otros bienes provenientes de la región del Báltico sur, 1700–1783", *Hispania. Revista Española de Historia* 76, no. 252 (2016), 129–158; "¿Cómo España trató de recuperar su poderío naval? Un acercamiento a las estrategias de la Marina Real sobre los suministros de materias primas forestales provenientes del báltico y Nueva España (1754–1795)", *Espacio Tiempo y Forma. Serie IV, Historia Moderna* 32 (2019), 73–102; "Direct Supplies of Timbers from the Southern Baltic Region for the Spanish Naval Departments during the Second Half of the 18th Century", *Studia Maritima* 33 (2020), 129–147; "El comercio de maderas del Báltico Sur en las estrategias de suministros de la Marina Real, 1714–1795", in Iván Valdez-Bubnov and Sergio Solbes Ferri y Pepijn Brandon (eds), *Redes empresariales*

comparison of these important regions for timber supply through the lens of the Spanish Royal Navy's development strategy was still lacking. Systematic work at Spanish, Mexican, Cuban, and Polish archives has shed light on trade relations, forest surveys, problems with transport, and use in naval construction of timber from the southern Baltic and New Spain, regions that presented very different geographical – economic – political – social contexts. However, owing to the determination of Spanish royal officials, Spain could ultimately meet its main target, which was to guarantee that its shipyards enjoyed a constant supply of timber from outside the Iberian Peninsula.

For this reason, this book cannot be easily slotted into a single field, such as environmental history[39] or maritime history.[40] Instead, it must be seen as a multifaceted attempt to cut across disciplinary boundaries,[41] to understand Spanish timber supply policies as a process that merged the environmental, economic, naval, political, geographic, military, and social histories of these two important regions (the southern Baltic and the viceroyalty of New Spain),

---

y administración estatal: la provisión de materiales estratégicos en el mundo hispánico durante el largo siglo XVIII (México: UNAM, 2020), 77–94; and "El transporte de maderas para los departamentos navales españoles en la segunda mitad del siglo XVIII", Studia Historica: Historia Moderna 43 (2021), 51–55; Torres Sánchez, Military Entrepreneurs, and "Los negocios con la armada. Suministros militares y política mercantilista en el siglo XVIII", in Iván Valdez-Bubnov and Sergio Solbes Ferri y Pepijn Brandon (eds), Redes empresariales y administración estatal: la provisión de materiales estratégicos en el mundo hispánico durante el largo siglo XVIII (México: UNAM, 2020), 49–76; Óscar Riezu Elizalde and Rafael Torres Sánchez, "¿En qué consistió el triunfo del Estado Forestal? Contractor State y los asentistas de madera del siglo XVIII", Studia Historica. Historia Moderna 43, no. 1 (2021), 195–226.

39   Alfred W. Crosby, Ecological Imperialism: The Biological Expansion of Europe, 900–1900 (Cambridge: Cambridge University Press, 1986); Edmund Burke and Kenneth Pomeranz (eds), The Environment and World History (Berkeley: University of California Press, 2009); Stephen Mosley, The Environment in World History (London: Routledge, 2010); John R. McNeill and Erin S. Mauldin, A Companion to Global Environmental History (Oxford: Wiley-Blackwell, 2012); Maohong Bao, "Environmental History and World History", Journal of Regional History 2, no. 1 (2018), 6–17.

40   Olaf Uwe Janzen, Merchant Organization and Maritime Trade in the North Atlantic: 1660–1815 (Newfoundland-St John: Liverpool University Press, 1998); Jan Glete, Warfare at Sea, 1500–1650: Maritime Conflicts and the Transformation of Europe (London: Routledge, 1999); Richard Harding, Seapower and Naval Warfare, 1650–1830 (London: Routledge, 1999); Carlos Martínez Shaw, "La historia marítima de los tiempos modernos. Una historia total del mar y sus orillas", Drassana 22 (2014), 36–64.

41   John Perlin, A Forest Journey: The Story of Wood and Civilization (Woodstock-Vermont: Countryman Press, 2005); Edward B. Barbier, Natural Resources and Economic Development (New York: Cambridge University Press, 2005); Paul Warde, Ecology, Economy and State Formation in Early Modern Germany (Cambridge: University of Cambridge, 2006).

their forested hinterlands, and harbours like Memel, Königsberg, Gdańsk/
Danzig, Szczecin/Stettin, Veracruz, Campeche, and Havana. In the 18th century, these places were intimately linked with the Spanish naval departments
of El Ferrol, Cádiz-La Carraca, and Cartagena through trade contracts and the
timber used to modernise the Royal Navy with the sponsorship of the state.

The timber trade that mobilised transport networks in North America, the
Caribbean, Spain, northern Europe, and the Baltic clearly illustrates the economic dimension of early globalisation, as it brought remote geographical
regions to interact closely.[42] It must be emphasised that a local product such
as European and American timber became, in this process, a global commod-
ity. The use of this wood in European shipbuilding contributed to colonisation
and early globalisation because the ships built in the royal shipyards were later
used to protect navigation routes across the Atlantic and the Pacific oceans.
Often, these ships, built with timber from Spain, the Baltic, New Spain, and
other places, were used to haul silver and other American products (gold,
sugar, tobacco, cochineal, and indigo) to Spain, and also took part in offensive
and defensive naval operations in time of war.

As noted, the book approaches Spanish timber supply policies from different points of view. For this reason, the monograph is divided into two parts
and comprises four chapters. The first part (Chapters 1 and 2), analyses timber
extraction and wood trade and transport from the southern Baltic to Spain. The
second part (Chapters 3 and 4) presents the Spanish Crown's efforts to organise
an efficient system to identify and harvest American timber from several provinces in the viceroyalty of New Spain, like Veracruz, Oaxaca, Campeche, Cuba,
Louisiana, and other forested regions in the Caribbean.

Chapter 1, entitled "Spanish Navy Wood Supplies from the Southern
Baltic and Riga", begins by presenting the policies devised by the Marquis of
Ensenada, one of the most powerful ministers during the reigns of Philip V
and Ferdinand VI, who faced the challenge of modernising the Spanish state
with administrative, fiscal, military, and naval reforms. As Minister of the Navy
and Indies, between 1743 and 1754 he continued the project to develop and
renovate the *Marina Real* begun by José Patiño y Rosales in 1717. He created
a network of consuls and spies to gather information about foreign armies,
navies, and military industries. During his time in office, the Navy adopted
the known system of *asientos*, which guaranteed public contracts with foreign
entrepreneurs associated with the Spanish commercial sector. The first major
contracts were signed in the 1750s to ensure the supply of timber and other

---

42   Bernd Hausberger, *Historia mínima de la globalización temprana* (México: Colegio de
      Mexico, 2018), 11–12.

raw materials, such as hemp and linen, for the naval departments of El Ferrol, Cádiz-La Carraca, and Cartagena. The chapter analyses the contracts proposed by different commercial houses – mostly Dutch, Irish, and Swedish – from 1750 to 1790. These contractors gave the Spanish Crown access to the Baltic markets, specifically the harbours between Szczecin/Stettin and Riga, which were able to supply excellent pine or fir masting, and pine – oak planks and beams. The chapter also examines attempts by Baltic merchants, for instance, Mathy and Schultz, from Gdańsk/Danzig, to compete with these businessmen, who held a virtual monopoly over timber exported to Spain.

Chapter 2 examines the hinterlands of three major regions in Prussia and the Polish-Lithuanian Commonwealth, which were connected by sea through navigable rivers. The first section addresses the trade in wood from Silesia and Kłodzko, which were floated down the Oder River to Szczecin/Stettin. This section also analyses the timber market at this harbour, to which wood was delivered from the jurisdiction of Pomerania. The second section of the chapter examines the flow of timber down the Vistula to Gdańsk/Danzig, a Hanseatic commercial hub in Poland and an important harbour for trade in cereal, wood, hemp, and other Baltic products. This section also includes information about the earliest legislation passed in the kingdoms of Prussia and Poland to protect the royal forests in Silesia, Kłodzko, and Kozienice. The last-but-one section analyses the forest economy of the influential magnate family of the Radziwiłłs and the commercial networks involved in the commercialisation of timber and other products in Riga, Memel, and Königsberg.[43] Finally, Chapter 2 examines Felipe Chone's trip to evaluate the timber floated down the Oder River to Szczecin/Stettin, as well as the Spanish Ambassador in the Kingdom of Poland, Count of Aranda's description of the Vistula timber trade, after taking a voyage downriver to Gdańsk/Danzig.

Chapter 3, "The Survey of New Spain's Woodlands", takes the research focus to the other side of the Atlantic, where, beginning in the 1760s, the Ministry of the Navy carried out several inspections of forests in the vast Spanish American empire in order to identify places where wood could be extracted. From the early 18th century onwards, royal projects had begun using timber from the region of Veracruz, in the viceroyalty of New Spain, which is traversed by several major rivers, including the Alvarado and the Coatzacoalcos. This region

---

43   This section was elaborated on and written by PhD students Karolina Juszczyk and Daniel Prusaczyk (who were scholarship holders on the project "The Role of Wood Supplies from the Southern Baltic Region and the Viceroyalty of New Spain in the Development of the Spanish Seaborne Empire in the Eighteenth Century" funded by the National Science Centre), whose work was supervised by the author of the present book.

was rich in tropical forests with good cedar, mahogany, and sabicú, while the mountains of Sierra Madre Oriental were covered in pine forests. The chapter analyses the strategies adopted by the Ministry of the Navy and the viceregal authorities to expedite the harvesting of wood for naval construction in Havana and Spain. Not all these inspections, undertaken by military and naval officers, resulted in wood-sourcing, but all of them yielded excellent geographical knowledge of the forested areas in the provinces of Veracruz and Oaxaca. The last section of the chapter examines timber-extracting projects in other regions of the viceroyalty, like the Usumacinta River, the Laguna de Términos, and the province of Huejutla.

Chapter 4 studies the results of wood logging in New Spain (provinces of Oaxaca, Veracruz, Tabasco, and Yucatán) but also in other regions of the Spanish Empire like Louisiana and the Caribbean, in the General Captaincy of Cuba and the Viceroyalty of New Granada. In New Spain – the region on which the main focus of this book lies – several *asientos* were signed with the support of the royal treasure. In the period 1784–1787, these contracts were granted to members of the military – commercial elite of Veracruz, who pledged to supply the *Marina Real* with *tozas* of mahogany and cedar and also prepared parts for ships-of-the-line and frigates. This project lasted only four years, not for lack of money or wood, but because of the constant problems with the transport of the timber from the river mouths to the shipyards in Cuba and Spain. The section on Louisiana examines pine-logging contracts with merchants from New Orleans to bring masting and planks to Havana. Finally, *asientos* granted in the Caribbean, specifically in the regions of Cumaná, the Magdalena River, and Cartagena de Indias, as well as the illegal felling and contraband of wood in Cuba's south-eastern forests are also analysed.

PART 1

∵

Spanish Navy Wood Supplies from the Southern Baltic and Riga

This chapter aims to explain the factors that shaped the decisions made by Spanish naval officials in the acquisition of planking and masting from northern Europe, especially the south Baltic region, which in the 18th century was divided between Prussia, the Polish-Lithuanian Commonwealth, and Russia. The region was covered by dense pine and oak forests, and had been used as a source of *tozas*, trunks, planks, and masting for naval construction in the Baltic harbours and elsewhere in Europe from the 15th century onwards.[1] The Baltic, therefore, became a key strategic area, as a source of large quantities of timber, for the Spanish Crown in the final years of Philip v's reign (1700–1746), a period that witnessed an arms race by the main maritime powers, including Great Britain, Holland, France, and, naturally, Spain. The size of war fleets became a decisive factor, which often determined the result of the struggle for colonial domination on several fronts worldwide.[2]

---

1   Czesław Biernat, *Statystyka obrotu towarowego Gdańska w latach 1651–1815* (Warsaw: PWN, 1962); Stanisław Gierszewski, *Statystyka żeglugi Gdańska w latach 1670–1815* (Warsaw: PAN, 1963); Artur Attman, *The Russian and Polish Markets in International Trade, 1500–1650* (Göteborg: Institute of Economic History of Gothenburg University, 1973); Artur Attman, *The Bullion Flow between Europe and the East, 1000–1750* (Göteborg: Kungl. vetenskaps-och vitterhets-samhället, 1981); Sven-Erik Åström, *From Tar to Timber: Studies in Northeast European Forest Exploitation and Foreign Trade 1600–1860* (Helsinki: Societas Scientiarum Fennica, 1988); W. Heeres and L. Noordegraaf, *From Dunkirk to Dantzig: Shipping and Trade in the North Sea and the Baltic, 1350–1850* (Hilversum: Verloren, 1988); Jerzy Trzoska, *Żegluga, handel i rzemiosło w Gdańsku w drugiej połowie XVII i XVIII wieku* (Gdańsk: Uniwersytet Gdański, 1989); Michael North, "The Export of Timber and Timber by Products from the Baltic Region to Western Europe, 1575–1775", in Michael North (ed.), *From the North Sea to the Baltic: Essays in Commercial, Monetary and Agrarian History, 1500–1800* (Norfolk: Variorum, 1996), 1–14; C.W. Pearson, *England's Timber Trade in the Last of the 17th and First of the 18th Century, More Especially with the Baltic Sea (1869)* (Whitefish: Kessinger, 2009); Rafal Reichert, "El comercio directo de maderas para la construcción naval española y de otros bienes provenientes de la región del Báltico sur, 1700–1783", *Hispania. Revista Española de Historia* 76, no. 252 (2016), 129–157; Rafal Reichert, "Direct Supplies of Timbers from the Southern Baltic Region for the Spanish Naval Departments during the Second Half of the 18th Century", *Studia Maritima* 33 (2020), 129–147.

2   Ralph Davis, *The Rise of the Shipping Industry in the Seventeenth and Eighteenth Centuries* (London: Macmillan, 1962); Nicholas A.M. Rodger, *The Wooden World: An Anatomy of the*

1   The Marquis of Ensenada and the Supply of Timber for the
       Spanish Navy

In 1743, Zenón de Somodevilla y Bengoechea—Marquis of Ensenada—was
appointed Secretary for the Navy and the Indies. He was a capable disciple
of José Patiño y Rosales, precursor of the modernization of the Spanish navy
in the opening decades of the 18th century. Using his knowledge of the organization of shipyards, arsenals, and naval departments in the Iberian Peninsula (Guarnizo, El Ferrol, Cádiz – La Carraca, and Cartagena) and the Indies
(Havana), the minister initiated a new policy for the supply of timber. Until his
time in office, most of the timber used for naval construction was sourced in
Spain and, in the case of the Havana shipyard, in Cuba.

In the 1740s, the Marquis of Ensenada was bestowed with wide powers, as,
in addition to the Secretary of the Navy and the Indies, he also led the Secretaries of State, Finances, and War. He used these sweeping powers to modernize
the army and navy during the War of Jenkins' Ear (1739–1748), in which Spain
fought Great Britain for the control of the American colonies. He implemented
a series of reforms to increase production in the Iberian Peninsula and trade
with the colonies, including the rationalization of taxation, more administrative control over the American colonies, and the reconstruction of the fleets of
the *Carrera de Indias*. These measures substantially increased public revenue,
and more money could be allocated to modernize the navy after the war with
Great Britain laid bare the operational problems of some of the king's ships-of-
the-line and frigates. When the war ended with the Treaty of Aix-la Chapelle
in 1748, Ensenada resumed the naval policies initiated by his predecessors—
Patiño y Rosales and Campillo y Cossío—and focused on accelerating the
administrative reform of the shipyards in the Iberian Peninsula and on creating a centralized administrative body to handle funding and naval supplies on

Georgian Navy (London: Collins, 1986); Jan Glete, *Navies and Nations: Warships, Navies and
State Building in Europe and America, 1500–1860*, vol. II (Stockholm: Almqvist & Wiksell
International, 1993); Jan Glete, *Warfare at Sea, 1500–1650: Maritime Conflicts and the Transfor-
mation of Europe* (London: Routledge, 1999); Richard Harding, *Seapower and Naval Warfare,
1650–1830* (London: Routledge, 1999); Olaf Uwe Janzen, *Merchant Organization and Maritime
Trade in the North Atlantic: 1660–1815* (Liverpool: Liverpool University Press, 1998); David
Ormrod, *The Rise of Commercial Empires: England and the Netherlands in the Age of Mercantilism, 1650–1770* (Cambridge: Cambridge University Press, 2003); Nicholas A.M. Rodger,
The Command of the Ocean: A Naval History of Britain, 1649–1815, vol. II (New York: Norton,
2004); Rafael Torres Sánchez, "Administración o asiento. La política estatal de suministros
militares en la monarquía española del siglo XVIII", *Studia Historica. Historia Moderna* 35
(2013), 159–199 and *Military Entrepreneurs and the Spanish Contractor State in the Eighteenth
Century* (Oxford: Oxford University Press, 2016).

a large scale. The minister aimed to keep pace with Great Britain by constantly increasing the size of the Spanish war fleet.[3]

Concerning timber, Ensenada encouraged the survey and assessment of the forest resources of Navarra, Basque Country, Cantabria, Asturias, and Galicia, on Spain's northern fringe; of Catalonia, Valencia, and Murcia, on the Mediterranean; and of Andalusia and Aragon, in the centre and south of the country. The surveys were undertaken between 1737 and 1739—before Ensenada's time—by navy officers, especially the frigate captain Juan Valdés y Castro, who was commissioned with mapping the most suitable species for naval construction in the Catalan, Andalusian, and Aragonese hill ranges. The systematic survey of the land allowed the Secretary of the Navy to locate interesting forest masses, for instance in Segura and Tortosa, where good pine and oak wood could be found in abundance, guaranteeing the supply of the naval departments in the Iberian Peninsula.[4]

Zenón de Somodevilla y Bengoechea, who knew the Spanish navy well, made an accurate diagnosis of supply problems, especially to do with the dispersion of felling contracts. As such, he began centralizing wood *asientos* through the Almirantazgo, for instance with Cantabria-based contractors Juan Bautista Donesteve and, especially, Juan de Ysla, who implemented a massive timber-sourcing programme in the woodlands of La Montaña in 1746 and 1749.[5] In addition, in order to prioritize the timber needs of the Navy, the *Ordenanza que su Majestad, manda observar para la Cría, Conservación, Plantíos y Corta de los Montes, con especialidad los que están inmediatos a la Mar, y Ríos Navegables. Método, y Reglas que en esta materia deben seguir los Intendentes de Marina, establecidos en los tres Departamentos de Cádiz, Ferrol y Cartagena* was published on 31 January 1748. This royal order put the woodland areas near the arsenals under the Navy's jurisdiction, and regulated the use of royal and private forests, giving priority to naval construction.[6] This strategy of limiting the felling of trees and reserving them for the Navy had negative consequences, because the wood business ceased being profitable for landowners and regions, which, until 1748, had greatly relied on this economic activity. In addition, landowners and fellers were paid set amounts for their wood, which

3   Iván Valdez-Bubnov, *Poder naval y modernización del Estado. Política de construcción naval española (siglos XVI–XVIII)* (Mexico: Universidad Nacional Autónoma de México, 2011), 271–273.
4   John T. Wing, *Roots of Empire. Forests and State Power in Early Modern Spain, c.1500–1750* (Leiden: Brill, 2015), 165–167.
5   Valdez-Bubnov, *Poder naval y modernización del Estado*, 260.
6   Alfredo Martínez González, *Las superintendencias de Montes y Plantíos (1574–1748)* (Valencia: Universidad de Sevilla, 2015), 333–357.

were below market prices. Felling trees for any purpose other than the state's naval needs had to be approved and overseen by naval officials.[7] It can be argued that the greatest beneficiary of this *Ordenanza* was Juan de Ysla, who, with a new contract signed in 1749, became the most powerful wood *asentista* in the 1750s.[8]

These measures, however, did not have the intended effect, and the problems continued piling up, notably the scarcity of wood for naval construction. This forced Ensenada to seek alternative solutions to push forth with the modernization of the armed forces. The Secretary of the Navy then began to think about sourcing the timber for the naval departments in the Baltic, the Mediterranean, and, to a lesser extent, Spanish America.[9]

Concerning Baltic timber, the first major contract was signed in March 1739 with Don Pedro de Aubert, *vecino* of Puerto Real, who agreed to sail to Riga to bring 12 cargoes of masting and planks to deliver in Cádiz, El Ferrol, and Cartagena. However, the beginning of the War of Jenkins' Ear in 1739 delayed the *asiento* for three years. The final cost of the contract, after adjusting the price of naval insurance and transport, was 42,952 pesos, 7 reales, and 15 maravedís de vellón.[10]

Ensenada, with the aid of Father Isla, Agustín Pablo de Ordeñana, and Facundo Mongrovejo,[11] created a network of informants and spies across Europe to report on commercial developments and opportunities in the continent. This network was supported by the Spanish consulates and embassies. The *Reglamento de sueldos de 1749*, which regulates the salaries of royal servants, includes informants in the main centres of power, such as Rome, Vienna, Paris, London, Saint Petersburg, and Berlin, and also the main commercial harbours,

---

7   Luis Urteaga, *La tierra esquilmada: las ideas sobre la conservación de la naturaleza en la cultura española del siglo XVIII* (Barcelona: Serbal, CSIC, 1987), 127–131.

8   Juan Miguel Castanedo Galán, "Un asiento singular de Juan Fernández de Isla. La fábrica de ocho navíos y la reforma de un astillero", in Carlos Martínez Shaw, *El derecho y el mar en la España moderna* (Granada: Universidad de Granada, 1995), 457–476.

9   Rafal Reichert, "¿Cómo España trató de recuperar su poderío naval? Un acercamiento a las estrategias de la marina real sobre los suministros de materias primas forestales provenientes del Báltico y Nueva España (1754–1795)", *Espacio, tiempo y forma. Serie IV, Historia moderna*, 32 (2019), 73–102.

10   Archivo General de Simancas (AGS), Secretaría del Despacho de la Marina (hereafter, Marina), legajo (hereafter, leg.), 618. Arboladura del Norte. Resultas del asiento de Aubert.

11   Cezary Taracha, "El Marqués de la Ensenada y los servicios secretos españoles en la época de Fernando VI", *Brocar* 25 (2001), 114–115.

including Stockholm, Gdańsk/Danzig, Copenhagen, Hamburg, Ostend, Nantes, Marseilles, Genoa, and Venice.[12]

Ensenada was very interested in the military industry and naval technology of other nations. Between 1746 and 1754, several spies, such as Jorge Juan, Pedro de Mora, and José Solano carried out secret missions in Britain, where, as well as gathering information, they recruited British naval builders for the Spanish naval departments, which began building their ships in the English style.[13] In addition, Antonio de Ulloa, Fernando de Ulloa, José de Azcarrati y Uztáriz, Salvador de Medina, and Enrique Enriqui travelled through France, Holland, Denmark, Prussia, Saxony, and Sweden to collect information about the metallurgical and naval industries, while José Manes, Dámaso Latre, and Agustín Hurtado went to England, Denmark, Sweden, Saxony, and Russia to collect information about the casting of artillery pieces and arms factories, as well as the organization of military academies and the technology of hydraulic machinery.[14]

Meanwhile, the Secretary of the Navy sent some of his officials to reach agreements with various European commercial houses, also in the Baltic, where different attempts were made to strike good deals for the Spanish navy. For instance, the Marquis of Puente-Fuerte, Spanish ambassador in Denmark, contacted Johan Frederik Classen, an influential arms and munitions supplier and adviser of King Frederik V (1746–1766), to negotiate the supply of timber and arms. In a report written in Copenhagen in 1753, the ambassador, who had begun collecting information about the Danish navy three years earlier, mentioned that the Danish king allowed a maximum of 1,500 masts from Fredrikshald to be sold abroad every year; however, the biggest masts were theoretically off limits, as these were reserved for the Danish shipyards. However, the Marquis of Puente-Fuerte continued, this law was not strictly enforced, so a way to buy large masts could be found. The ambassador attached a detailed list of measures and prices, comparing the evolution of the price of timber since 1751; prices had gone up by 24 per cent in barely two years, because of the

---

12   Archivo Histórico Nacional (AHN), Secretaría del Estado (hereafter, Estado), leg. 3439 expediente (hereafter, exp.), 33. Reglamento original de sueldos del año de 1749.

13   The main British builders hired by the Spanish navy were Richard Rooth (El Ferrol), David Howell (Guarnizo), Edward Bryant (Cartagena), and Matthew Mullan and Almond Hill (Cádiz – La Carraca). Antonio Lafuente and José Luis Peset, "Política científica y espionaje industrial en los viajes de Jorge Juan y Antonio de Ulloa", *Melanges de la Casa de Velázquez* 17 (1981), 233–262; Valdez-Bubnov, *Poder naval y modernización del Estado*, 276.

14   Taracha, "El Marqués de la Ensenada", 117.

large demand posed by the construction of a new neighbourhood in Copenhagen, Amalienborg, under the direction of the royal architect, Nicolai Eigtved.[15] Concerning the construction of Danish ships-of-the-line, the Marquis of Puente-Fuerte wrote that:

> timber for construction is not overabundant in the Danish forests. In Holstein there is more, and the locals use it a little for building commercial ships, and the rest is felled and bought by the Dutch. However, the timber that H. M. uses for his Royal Fleet is brought from Pomerania; except for pine boards and some masts, which he brings from Norway, and the largest ones from Riga.[16]

Indeed, not the whole Baltic region was endowed with forests of a sufficient quality to supply wood suitable for naval construction. According to Sven Åström,[17] Jan Glete,[18] Michael North,[19] David Ormrod,[20] and more recently Rafal Reichert,[21] most of the demand posed by maritime powers focused on the southern Baltic, through the important harbours of Szczecin/Stettin, Gdańsk/Danzig, Königsberg (modern Kaliningrad), Memel (modern Kláipeda) and, farther east, Riga. These major harbours, assisted by several important river routes—Oder, Vistula, Pregolia, Niemen, and Daugava—became strategic

15 AHN, Estado, leg. 3397, exp. 5. Informes de la Marina de Guerra de Dinamarca.

16 la madera propia para la construcción no es muy común en los bosques de Dinamarca. En Holstein se haya con más abundancia y sus habitantes emplean una pequeña parte a la fábrica de sus navíos para comercio y la demás la sacan y compran los holandeses. No obstante, la madera que S.M. Dinamarqués necesita para su Real Flota, la haga venir de Pomerania; excepto las tablas de pino y algunos mástiles que saca de Noruega y los grandes de Riga. (AHN, Estado, leg. 3397, exp. 5. Informes de la Marina de Guerra de Dinamarca).

17 Sven-Erik Åström, "English Timber Imports from Northern Europe in the Eighteenth Century", *Scandinavian Economic History Review* 18 (1970), 12–32.

18 Glete, *Navies and Nations*.

19 North, *From the North Sea to the Baltic*.

20 David Ormrod, "Institutions and the Environment: Shipping Movements in the North Sea and Baltic Zone, 1650–1800", in Richard W. Unger (ed.), *Shipping Efficiency and Economic Growth, 1350–1800* (Leiden: Brill, 2011), 135–166.

21 Reichert, "El comercio directo de maderas", 129–158; Reichert, "¿Cómo España trató de recuperar su poderío naval?", 73–102; Reichert, "Direct Supplies of Timbers", 129–147; Reichert, "El comercio de maderas del Báltico Sur en las estrategias de suministros de la Marina Real, 1714–1795", in Iván Valdez-Bubnov, Sergio Solbes Ferri, and Pepijn Brandon (eds.), *Redes empresariales y administración estatal: la provisión de materiales estratégicos en el mundo hispánico durante el largo siglo XVIII* (Mexico: Universidad Nacional Autónoma de México, 2020), 77–94.

hubs in the supply of cereal, vegetal fibres (hemp and flax), and, naturally, timber (firewood, staves, and naval parts), brought from interior Prussia, Saxony, the Polish-Lithuanian Commonwealth, Eastern Prussia, and Russia.

Ensenada was very interested in the southern Baltic region and Riga as a source of raw materials. Between 1746 and 1748, he sent his loyal associate Jacinto Ferrero Fieschi y de Saboya, best known as Count of Bena, as minister plenipotentiary to the court of August III Wettin, King of Saxony and the Polish-Lithuanian Commonwealth. During his visit to Dresden and Warsaw, Ferrero sought to finalize a commercial agreement with the Polish-Lithuanian union, which could supply flexible and suitable timber for masting.[22] He fulfilled his mission in February 1749. Before returning to Spain, he travelled to Gdańsk/Danzig, where he purchased two cargoes for the naval department in Cádiz, one of masting and another one of planking and other naval parts, from the commercial house led by Ignacio Jacinto Mathy,[23] a French company based in Gdańsk/Danzig from the late 17th century. In the event, when the cargoes arrived in Cádiz, the navy only bought the masts, while the boards were sold to private merchants in Cádiz, but at a loss.[24] However, the French representatives in Cádiz, Christian Sentrup and Mathy's son, Antón Jacinto, reported that the Spanish navy officers were in awe of the wood, as they claimed "never to have seen such good timber for masting".[25] Interestingly, the timber was hauled by one of the most powerful merchants of Gdańsk/Danzig, Juan Felipe Schultz, who owned a ten-ship fleet and in 1740 became *asentista* of the masts and planks needed to repair the king's ships and galleys in the French departments of Toulon and Marseilles.[26]

Before going on to describe specific *asientos* signed by the Spanish Crown to purchase south Baltic timber, it is worth examining geographic information

---

22   Didier Ozanam, *Les diplomates espagnols du XVIIIe siècle. Introduction et répertoire biographique (1700–1808)* (Madrid: Casa de Velázquez, 1998), 189.

23   Between 1704 and 1757, the Mathy or Matthy family represented French commercial interests in Gdańsk/Danzig, and some of its members acted as legal representatives of the French king. In 1715, they also began collaborating with the French navy, when Luis Mathy was appointed navy commissar in Gdańsk/Danzig. Edmund Cieślak, *Francuska placówka konsularna w Gdańsku w XVIII wieku. Status prawny, zadania, działalność* (Krakow: Polska Akademia Umiejętności, 1999), 204.

24   Ten large masts, between 35 and 41.5 cubits long, and 66 small ones, between 33 and 21 cubits long. AGS, Marina, leg. 612. Factura de arboladura.

25   "nunca antes vieron un cargamento de tan bella y buena madera de arboladura". Ibid. Asiento de arboladura.

26   In the 1740s, the Polish King August III granted Juan Felipe Schultz a contract to trade wood from the royal forests of the Polish-Lithuanian Commonwealth with Spain and France. Trzoska, *Żegluga, handel i rzemiosło w Gdańsku*, 167.

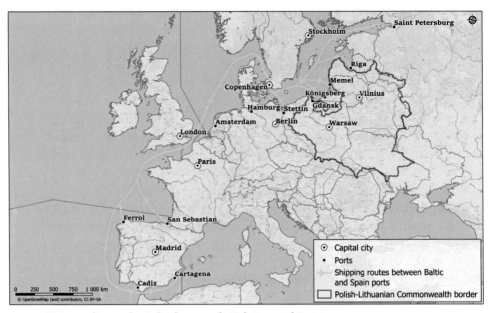

MAP 1   Main naval routes for timber between the Baltic sea and Spain
          SOURCE: LOOK4GIS-LUKASZ BRYLAK BASED ON QGIS SOFTWARE

system (GIS) maps that display the naval routes followed by the Dutch, English, French, Scandinavian, and Baltic merchants to bring the timber to the naval departments of El Ferrol, Cádiz, Cartagena, and to the port of San Sebastián (see Map 1). Map 2 shows the main European rivers with special focus on those that run through the territory of the Polish-Lithuanian Commonwealth before its first partition by Prussia, Russia, and Austria in 1772 (see also Figure 2). This is essential, because these rivers were the main waterways along which timber, but also other goods, were hauled, connecting the interior of the Commonwealth with the main southern Baltic harbours (see Map 3).

2   Mathy-Schultz Versus Gil de Meester

In 1749, after sending this initial cargo and the positive response to the wood, Mathy, with Schultz's assistance, prepared a two-year offer to supply the departments of Cádiz, El Ferrol, and Cartagena with 620 masts from Riga each year.[27] This proposal nearly coincided in time with the contract presented, in 1750, by

---

27   AGS, Marina, leg. 612. Asiento de arboladura.

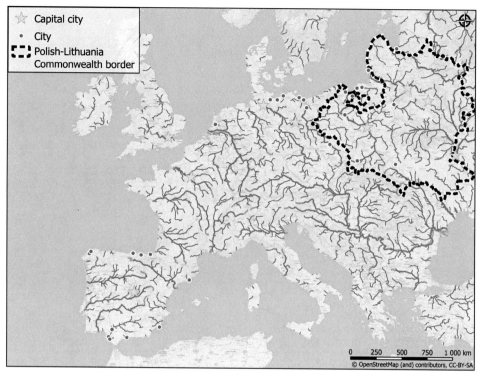

MAP 2   European rivers and forests and the borders of the Polish-Lithuanian Commonwealth before 1772
SOURCE: LOOK4GIS-LUKASZ BRYLAK BASED ON QGIS SOFTWARE

the Lisbon-based commercial house led by the Dutch Juan and Daniel Gil de Meester. These two brothers had worked with the Spanish navy since 1744 when they were awarded a contract to supply pitch and wood from Tortosa.[28] Ignacio Jacinto Mathy tried to prove that his offer was better than Gil de Meesters' with the following arguments:

- His commission was 3 per cent, which, according to him, no other merchant offered, as everyone else sought to make a greater profit.
- His masts were of top quality, which, according to Mathy, Gil de Meesters was in no position to offer, because the *asiento* presented by the Dutch

---

28   Rafael Torres Sánchez, "Los negocios con la armada. Suministros militares y política mercantilista en el siglo XVIII", in Iván Valdez-Bubnov, Sergio Solbes Ferri, and Pepijn Brandon (eds.), *Redes empresariales y administración estatal: la provisión de materiales estratégicos en el mundo hispánico durante el largo siglo XVIII* (Mexico: Universidad Nacional Autónoma de México, 2020), 52–53.

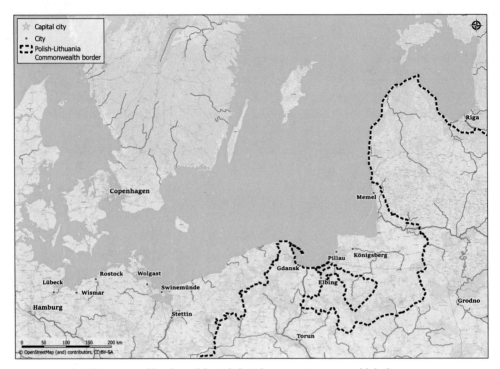

MAP 3   South Baltic ports and borders of the Polish-Lithuanian Commonwealth before 1772
         SOURCE: LOOK4GIS-LUKASZ BRYLAK BASED ON QGIS SOFTWARE

mentioned "masting from the North", which was a vague term that freed the merchant's hands to source the wood from anywhere in the Baltic region.

– Mathy committed to deliver Spain only the best-quality masts from Riga.

– Finally, Mathy suggested that the Gil de Meesters could, with their enormous resources, bribe the Spanish navy officials sent to oversee the measures and quality of the goods.[29]

In his proposal, Mathy asked for the masts to be reviewed and measured in Riga to ensure that they were the right size, came from healthy trees, and did not have major knots that undermined their resistance.[30] In contrast, the *asiento* signed in September 1750 with Eugenio Mena, representative of the Gil de Meester house, for 916 masts from the Baltic, made no specific mention of the origin of the trees and included only a general reference to "good quality".[31]

---

29   AGS, Marina, leg. 612. Asiento de arboladura.
30   Ibid.
31   AGS, Marina, leg. 787. Asiento de arboladura del Norte para arsenales de El Ferrol, Cádiz y
      Cartagena.

FIGURE 2   The Polish-Lithuanian Commonwealth before and after the 1772 partition
SOURCE: ARCHIWUM GŁÓWNE AKT DAWNYCH, 403 ZBIÓR GEOGRAFICZNY
STANISŁAWA AUGUSTA, SYGNATURA 43, KARTA 33

A summary of the state of pine planking for three ships under construction in
Cartagena, dated 27 March 1753, mentions that the wood came from Prussia
(11,066 parts of different lengths and thicknesses),[32] Riga (6,300 parts), Norway
(3,600 pieces of white planking),[33] and Tortosa (3,250 parts).[34] A report signed
by Francisco Barrero Peláez about the arrival of ships in the department of
Cartagena mentions that a Dutch fluyt named *Daniel*, captained by Dirk Dirk,

---

32   According to custom registers in the Sund, this probably meant East Prussia, the harbours
of Königsberg and Memel. Reichert, "El comercio directo de maderas", 129–158.

33   Probably from the region of Kopervik. AGS, Marina, leg. 612. Etat des mats et planches du
Nord.

34   Ibid. Estado que comprende la tablonería de pino que según reglamento corresponde a la
entera construcción de los tres navíos.

arrived on 28 January 1753 from Gdańsk/Danzig loaded with pine masting on behalf of Don Juan and Don Daniel Gil de Meester; the cargo was unloaded for inspection.[35] Therefore, Mathy's appreciations were accurate, as it was clear that the Dutch merchants were acquiring their wood in various Baltic harbours, as they could not source it all from Riga.

Despite Mathy's efforts to show that the Dutch were sourcing the wood from all over the Baltic instead of only Riga, his offer was rejected and the Gil de Meesters accepted, first for one year (1751), during which the Dutch delivered 14 cargoes of northern planking and masting to the department of Cádiz, for 2,968,463 reales de vellón; just one to Cartagena, for 187,457 reales de vellón; and eight to El Ferrol, for 1,040,810 reales de vellón.[36] In addition to the 916 masts, the Gil de Meesters' first *asiento* included boards for 19 70-gun ships (12 of which were to be built in El Ferrol, four in Cádiz, and three in Cartagena); this project requested 78,000 pine boards of different length and thickness, of which 55,100 were to be bought in East Prussia and Norway (see Table 1). The naval architect Richard Rooth, who was in charge of operations in El Ferrol, estimated that each ship required around 4,000 boards, and this estimate was also applied to the ships being built in Cádiz and Cartagena.[37]

TABLE 1   Estimate of the pine planking needed to build 19 70-gun ships in the departments of El Ferrol, Cádiz, and Cartagena

Pine boards	El Ferrol	Cádiz	Cartagena	Total
3.5 inches thick and 24 cubits long	1,800	600	450	2,850
3.5 and 2.5 cubits	1,800	600	450	2,850
3 and 21.5 cubits from Prussia	3,600	1,200	900	5,700
3 and 19 cubits	3,600	1,200	900	5,700
2.5 and 19 cubits from Prussia	3,600	1,200	900	5,700
2.5 and 16 cubits from Prussia	3,600	1,200	900	5,700
2.25 and 5 cubits	6,000	2,000	1,500	9,500
1.5 and 6.5 cubits from Norway	6,000	2,000	1,500	9,500
1.5 and 5 cubits from Norway	18,000	6,000	4,500	28,500
Total	48,000	16,000	12,000	76,000

SOURCE: AGS, MARINA, LEG. 612. RELACIÓN DE TABLAZÓN DE PINO QUE DEBE PROVEER EN EL AÑO 1751 EL ACTUAL ASENTISTA

35   Ibid. Entrada de buque holandés al puerto de Cartagena.
36   AGS, Marina, leg. 608. Estado de la cuenta que tiene Arboladura del Norte.
37   Ibid., leg. 612. Razón de la madera de pino para navíos de 70 cañones.

These efficient logistics convinced the admiralty to prorogue the contract with the Dutch company until 1759. In the new *asiento*, the navy demanded the masting to be bought in Riga, to be of the best quality, and to stick to the measurements specified in the contract. The Secretary of the Navy requested Don Antonio de Ulloa, a famous navy captain, to assess the agreement. Ulloa, comparing the prices offered by the Meesters with those being paid in Sweden, determined that the contract was highly advantageous to the *Real Hacienda*, and "it will be more so in the future, given the scarcity of masting that the North is beginning to suffer and the constant purchases made by the French, the English [*sic*], and the Dutch".[38]

However, a new Spanish consul, Luis Perrot, arrived in Gdańsk/Danzig in 1752 with the order to promote trade between Spain and that port, which linked the Polish, Prussian, and Russian commercial networks with Western Europe. The Bourbon government aimed to import wheat, barley, oats, wax, fat, hides, and hemp for cables, as well as sails, masts, and other materials for naval construction. In exchange, Spain offered wine, salt, oils, and dry fruit. Immediately after arrival, Luis Perrot, reached out to the commercial circles in Gdańsk/Danzig, and soon met Mathy and Schultz.[39] In 1753, he made a proposal to Spain presented by Juan Felipe Schultz,[40] concerning the direct supply of Polish and Livonian timber to the Spanish arsenals. The proposal argued that, to that date, the Gdańsk/Danzig merchants had worked for the Gil de Meesters, and that it was them who originally supplied the timber, which the Dutch later sold to the Spanish shipyards; the Schultzs, therefore, wanted to dispense with the middleman. This, however, never came to pass, and the supply of masting and planking remained in the hands of the Gil de Meester brothers.[41]

Apparently, the service lent by the Lisbon-based Dutch company finally met Ensenada's target of creating a unified system for the transport and distribution of timber to the naval departments. In addition, the contract allowed the

---

38   "mucha mayor en lo sucesivo por la escasez de arboladura que se va experimentando en el Norte y la continuada compra que hacen Franceses, Ingleses y Holandeses". AGS, Marina, leg. 612. Asiento de arboladura del norte celebrado con los Meester por ocho años.

39   Adolf Poschman, "El consulado español en Danzig desde 1752 hasta 1773", *Revista de Archivos, Bibliotecas y Museos* 4–6 (1919), 210–212.

40   Juan Felipe Schultz died in 1754, leaving a seven-ship fleet to his heirs. Edmund Cieślak, "Gdański projekt kasy wykupu marynarzy z rąk piratów z połowy XVIII wieku", *Przegląd Historyczny* 51 (1960), 43–44.

41   Trzoska, *Żegluga, handel i rzemiosło w Gdańsku*, 168; Torres Sánchez, "Los negocios con la armada", 61.

TABLE 2   Direct trade between Gdańsk/Danzig and Spain in the 1750s, according to the customs office of the Sund (Denmark)

Year	1751	1752	1753	1754	1755	1756	1757	1758	1759	1760
Ships loaded with timber	2	9	12	13	1	4	6	2	3	1
Total ships	16	39	57	18	10	20	16	9	14	11

SOURCE: REICHERT, "EL COMERCIO DIRECTO DE MADERAS" Y REGISTROS DIGITALES DE SUND, HTTP://WWW.SOUNDTOLL.NL/INDEX.PHP/EN/WELKOM. N. DE MICROFILMS: 245, 246, 247, 248

supply to be rationally organized during the 1750s, which was very important for the renovation of the *Marina Real*. Although the Mathy commercial house was not awarded the *asiento*, its negotiation was instrumental in creating a direct link between Spain and the southern Baltic, and the 1750s witnessed a continued increase in the flow of Baltic timber for naval construction until Spain joined the Seven Years' War in January 1762 (see Table 2).[42]

3   Baltic Timber and the *Asiento* of the Gil de Meester Brothers (1751–1760)

By signing the *asientos* with the Meester brothers, the Secretary of the Navy ensured a constant supply of masting and planking from the Baltic. If the transport of the wood was relatively unproblematic, as confirmed by the ledgers of naval department accountants, who kept a detailed record of sea traffic in their stations,[43] the measures of the parts served were much less straightforward, because parts of major and minor masting and planks often did not match the dimensions established in the contract. For instance, on 5 January 1752, Francisco Barrero Peláez, intendent of the Cartagena department, wrote a letter to Ensenada in which he pointed out that Don Eugenio de Mena, the Meesters' representative, had delivered a cargo with nine masts too many (the order was for 178), but that none of them could be used for either major masts or foremasts. The intendent added that something similar had happened with the cargo that

---

42   Reichert, "El comercio directo de maderas", 129–157.
43   Rafal Reichert, "El transporte de maderas para los departamentos navales españoles en la segunda mitad del siglo XVIII", *Studia Historica: Historia Moderna* 43 (2021), 51–55.

arrived in Cartagena on 21 December 1751. In his report, he emphasized that the parts delivered had been measured carefully, that most did not meet the thickness and length required by contract, and that they could only be used in frigates. In the final line of his letter, Francisco Barrero Peláez asked the Secretary of the Navy to "tell the *asentista* [to] send the [masts] required for 70-gun ships as soon as possible, so that when they are needed we do not find that they are lacking".[44]

One of the ships under construction at the time in Cartagena was the *Septentrión*. In order not to delay the works, Don Eugenio de Mena decided to send the Dutch fluyt *Concordia*, captained by Haytie Hering, just arrived in Cádiz with 47 northern masts, to the Levantine department. However, the masting master at Cartagena, Pedro de Hordeñana, after revising the masts, reported that "although the pieces delivered are not entirely in conformity with the contract because they are bigger or smaller than required, they have been combined with others so that they can be used".[45] The same happened with the boards for the *Septentrión*, the lengths of which did not comply with the contract, so it was decided "to use other [timber] of the same thickness but shorter, delivered by the skipper [Hering] for other purposes".[46] This problem sprang from the fact that the ship had been designed in 1740, following Ciprián Autrán's specifications, but was eventually built according to the British system under the supervision of Edward Bryant.[47] On the other hand, the measurements in cubits and inches used by the Dutch, British, and Spanish—all the nations involved in the Gil de Meesters' contract—were different.[48] The *asentistas* were thus required to be mindful of the measurements, turning the Baltic units into the Spanish, but this was not always done because of the simple fact that the Meesters had no direct representatives in Riga, Königsberg, Memel, and Gdańsk/Danzig, but relied on the cooperation of local merchants,

44   "mandara al citado asentista [que] envíe aquí los [palos] que correspondan a navíos de 70 cañones con la posible brevedad a fin de que a su tiempo no se experimente su falta". AGS, Marina, leg. 612. Acompaña relación de la arboladura que falta proveer por Mena.

45   "aunque las piezas recibidas no se conforman precisamente con las de la contrata en sus medidas por sobrar o faltar alguna cosa indiferentemente ya en el grueso ya en el largo, se han combinado con las de las clases que se figuran por adaptarse y así puedan ser útiles". AGS, Marina, leg. 612. Nota de los árboles del Norte.

46   "con otra [madera] de los mismos gruesos pero más corta en sus respectivas clases que entregó el transportista [Hering] para servicio de las obras del proyecto y otros usos". AGS, Marina, leg. 612. Nota de la tablazón de pino.

47   Valdez-Bubnov, *Poder naval y modernización del Estado*, 284–285.

48   "All the inches in the diameters are reduced to hands, which is the measurement used by masting masters, by multiplying the inches by 3 and dividing by 8, which is the Dutch system". AGS, Marina, leg. 612. Medidas por codo inglés para largos y gruesos de los baupreses.

who—occasionally—loaded whatever timber they had in stock.[49] Because of these issues, the *Septentrión*, finished in 1751, was nearly 5 cubits shorter than designed, and its ordnance had to be reduced from 70 to 68 guns.[50]

In January 1752, the intendent of the navy in El Ferrol, Bernardino Freyre, made the report about the northern masting and planking sent by the Meesters to his department in the previous year. The document accounts for 363 masts, but not all of them met the conditions specified in the contract. For instance, the shipyard had ordered 24 masts from 9¾ to 9 hands thick and from 32 to 49 cubits long, but 52 were delivered; in addition, 47 masts, from 6¾ to 6 hands thick and from 24 to 39½ cubits long were delivered without having been ordered, and had to be transferred; finally, 171 masts from 4¾ to 2 hands thick and from 16 to 31½ cubits long were delivered, but this fell 97 masts short of the order. Of the 363 masts delivered, 79 failed to meet the measures set out in the contract, and 165 more were altogether missing. The situation with planking was not much better, Gil de Meesters delivered 23,486 boards of different sizes, but the order was for 30,334 pieces; 5,820 of the boards delivered failed to meet the contract's specifications.[51]

These discrepancies clearly indicate that the monitoring system used to oversee the deliveries was not working efficiently. It has been noted that the problems began in the southern Baltic, where the Meesters had no representatives,[52] and where there were no Spanish navy officers either, although it was these who sent yearly memoranda with estimates of the northern timber that was to be needed by naval departments.[53] These estimates, made by the intendents at the end of each year, listed the masting and planking required and their exact precise measurements. These annual estimates were to become a rule for all future *asientos* concerning Baltic timber. It should be noted that, in the 1750s, the Spanish Crown had a single consul in the southern Baltic, specifically in Gdańsk/Danzig, and according to Luis Perrot's report, he could

---

49   AGS, Marina, leg. 616. Medidas exactas de los pies y palmos con que se miden en el Báltico
      las maderas de construcción.

50   El Ferrol and Cádiz – La Carraca were also forced to shorten the keel and reduce the ord-
      nance aboard the ships under construction. As a result, not a single two-decker with 70
      guns was built during the 1750s; the ships built during this period had 68, 60, or 58 guns.
      See Valdez-Bubnov, *Poder naval y modernización del Estado*, 285–289.

51   AGS, Marina, leg. 612. Estado de arboladura y tablonería del Norte.

52   The Meesters bought the missing timber in Amsterdam, for instance in 1753, when they
      bought 120 masts there. AGS, Marina, leg. 612. Factura de las medidas que tienen 269 árbo-
      les comprados en el Norte.

53   AGS, Marina, leg. 612. Relación de la arboladura de Riga para el departamento de El Ferrol.

supervise some of the cargoes being prepared for the navy.[54] The constant complaints filed by navy officers were finally addressed in a new *asiento* signed with Gerónimo de Retortillo in 1760. The *asentista* explained in a letter that, following said complaints, the harbour of Riga had published a set of ordinances concerning the "observations related to the thickness and length of trees growing in Poland, Ukraine, Lithuania, and other provinces".[55] The new contract introduced standardized measurements; for instance, it was agreed that masts that were 10 hands thick were to be from 40 to 43 cubits long. More importantly, the authorities in Riga committed to appoint sworn inspectors to oversee that "the masts are admissible, in term of both size and quality";[56] conversely, the harbour officials ensured that, once the inspectors had given their *placet* to a piece, no claims could be filed.[57]

The last factor to be taken into account to understand the issue of measurements was the competition between Spain and other European powers, especially Great Britain, Holland, and France, to acquire these raw materials, which caused supply shortages in the thickest and longest pieces from the 1740s onwards. For instance, one of the 12 68-gun two-deckers built in El Ferrol in the 1750s required a total of 113 masts. The hardest pieces to find were the main masts (three per ship), main yard (one), foremasts (three), mizzens (one), bowsprits (three), and *gimelgas* of foremast and bowsprit (three);[58] that is, the massive construction programme undertaken by all naval powers in the second half of the 18th century made for a massive aggregate demand and caused supply shortages of the thickest and longest masts, which could only be found in southern Baltic harbours (Memel, Königsberg, Gdańsk/Danzig); Norwegian harbours could only supply small and medium masting.[59]

Table 3 presents the price list corresponding to the Gil de Meesters' first contract in 1751, which shows just how expensive thick and long masts could be. This is also confirmed by the registers of the ships bringing the masting to the

---

54   Jesús Pradells Nadal, "Los cónsules españoles del siglo XVIII. Caracteres profesionales y vida cotidiana", *Revista de Historia Moderna* 10 (1991), 251; Dorota Nowak, "Konsulat hiszpański w Gdańsku w drugiej połowie XVIII wieku. Kilka uwag o możliwościach badawczych", *Roczniki Humanistyczne* 58 (2010), 127–128.

55   "Observaciones hechas por la práctica de largo y grueso que comúnmente tienen los árboles que se crían en montes de Polonia, Ucrania, Lituania y otras provincias". AGS, Marina, leg. 616. Instancias del asentista de arboladura don Gerónimo de Retortillo.

56   "si los palos son admisibles tanto por sus proporciones, como por la calidad de la madera".

57   AGS, Marina, leg. 616. Instancias del asentista de arboladura don Gerónimo de Retortillo.

58   Ibid. Estado del número y dimensiones de los palos del Norte que se necesitan para la arboladura de un navío de 68 cañones.

59   Åström, "English Timber Imports from Northern Europe in the Eighteenth Century", 16.

TABLE 3   Examples of prices of different kinds of mast in the Gil de Meesters' *asiento* (1751)

Hands thick	Cubits long	Reales de vellón
11½	40	9,000
10½	47	7,500
10½	36	7,500
9½	42	5,500
8½	46	3,000
7½	41	1,700
7½	34	1,700
5½	39	600

SOURCE: AGS, MARINA, LEG. 612. DON EUGENIO DE MENA APODERADO DE JUAN Y
DANIEL GIL DE MEESTER SE OBLIGA A PROVEER [...] LA ARBOLADURA DEL NORTE DE
BUENA CALIDAD

TABLE 4   Number of ships-of-the-line built in the royal shipyards between 1751 and 1756

Cartagena	Cádiz	El Ferrol	Guarnizo
3	3	13	4

SOURCE: IVÁN VALDEZ-BUBNOV, *PODER NAVAL Y MODERNIZACIÓN DEL ESTADO. POLÍTICA
DE CONSTRUCCIÓN NAVAL ESPAÑOLA (SIGLOS XVI–XVIII)* (MÉXICO: UNIVERSIDAD
NACIONAL AUTÓNOMA DE MÉXICO, 2011), 297–298

Spanish naval departments. For instance, the Dutch *Willemina*, under skipper
Juan Breevilt, arrived in El Ferrol in December 1752, with 70 masts from Riga,
31 of which were between 11½ and 8½ hands thick and 48¾ and 31¼ cubits
long; or the Dutch fluyt *Westermer*, captained by Claas Douvezsei, who arrived
also in El Ferrol from Riga in November 1753, with 73 trees, 58 of which were
between 11½ and 8½ hands thick and 48 and 30 cubits long, and the remaining
15 were between 6¾ and 4 hands thick and 36 and 26 cubits long.[60]

These and other examples show that, despite the difficulties in finding
masting for mainmasts, foremasts, mizzens, and bowsprits because of the
enormous demand posed by all naval powers, the Spanish navy was receiving a

---

60   AGS, Marina, leg. 612. Estado de la arboladura del Norte.

substantial number of large good-quality masts. In fact, as Iván Valdez-Bubnov points out as many as 23 ships-of-the-line and a dozen smaller ships were built in the shipyards of Cartagena, Cádiz, El Ferrol, and Guarnizo between 1751 and 1756, that is, in the first five years of the Gil de Meesters' *asiento*.[61]

Rafael Torres Sánchez reports that an important change was introduced in the mechanics of wood supply in 1760. This change sprang from the rivalries of potential contractors, which gave the navy a bargaining advantage. The competition between the Dutch company of the Gil de Meesters (the old monopolist that controlled the supply of naval departments in the 1750s) and the commercial house of Raimundo[62] and Miguel Soto (two Irishmen based in Cádiz, with links with the main Spanish merchants) resulted in considerable savings for the state's treasure.[63] In addition, both competitors offered to bring the best timber in the North, especially from Riga. This ended the monopoly of the Gil de Meesters, and the navy signed a two-year contract with the Soto. However, in June 1760 a new *asiento* came up for tender to cover the six following years (1761–1766), to which the Count of Clonard, the Gil de Meesters, and Simón de Aragorri y Olavide (an influential Spanish merchant who used his political connections to be awarded public supply contracts by the Bourbon Crown) concurred. According to the evaluation of the tenders, the best offer was the latter, because it priced the timber 36 per cent below what the Gil de Meesters had been charging in 1751.[64] The minister of finance, the Marquis of Esquilache, who went with this evaluation, soon saw the financial advantages of the proposal and the opportunity to Hispanicize the wood *asiento*, as two of the three commercial houses with contracts operated in Spain and had close links with Spanish commercial networks, such as the Cinco Gremios Mayores.[65]

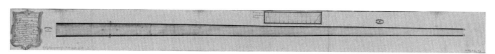

FIGURE 3   Pine mast from Riga
SOURCE: ESPAÑA. MINISTERIO DE CULTURA Y DEPORTE. ARCHIVO GENERAL DE SIMANCAS, MAPAS, PLANOS Y DIBUJOS, 23, 076

---

61   Valdez-Bubnov, *Poder naval y modernización del Estado*, 297–298.
62   Also known as 1st Count of Clonard.
63   The Sotos offered prices 25 per cent below what the Meesters had been charging in the 1750s.
64   AGS, Marina, leg. 787. Asiento de arboladuras del Norte por 6 años a comenzar en el de 1761.
65   Torres Sánchez, "Los negocios con la armada", 64–67.

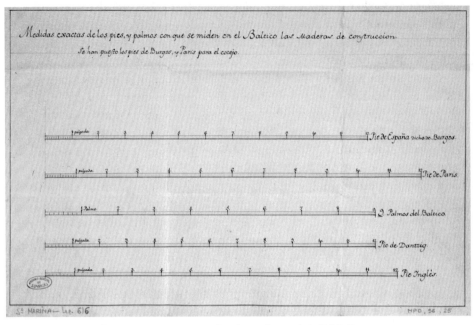

FIGURE 4   Foot and span measurements used in the timber trade in Baltic Ports
SOURCE: ESPAÑA. MINISTERIO DE CULTURA Y DEPORTE. ARCHIVO GENERAL DE
SIMANCAS, MAPAS, PLANOS Y DIBUJOS, 56, 025

4   Retortillo and Aragorri's Wood *Asiento* (1761–1766)

Aragorri's *asiento* during the period 1761–1766, which was implemented in practice by his representative Gerónimo de Retortillo, had a second aim: to increase
the Spanish commercial presence in the Baltic ports. For this reason, both the
number of consuls and the visits of royal officials to cities such as Saint Petersburg, Riga, Königsberg, and Gdańsk/Danzig increased in the 1760s.[66] Concerning Gdańsk/Danzig, the Spanish ambassador to Poland, the Count of Aranda,
made a trip from Warsaw to convince the merchants at Gdańsk/Danzig to
conduct more direct deals with their Spanish counterparts. On his return to

---

66   Pradells-Nadal, "Los cónsules españoles del siglo XVIII", 221–262; Cezary Taracha, "Jeszcze
o gdańskiej misji Pedra Arandy w 1761 roku", *Rocznik Gdański* 56, no. 2 (1996), 17–21; Klemens Kaps, "Trade Connections between Eastern European Regions and the Spanish
Atlantic during the Eighteenth Century", in Katja Castryck-Naumann, *Transregional Connections in the History of East-Central Europe* (Berlin: De Gruyter, 2021), 217–258; Cezary
Taracha, *Spies and Diplomats: Spanish Intelligence Service in the Eighteenth Century*
(Frankfurt: Peter Lang, 2021).

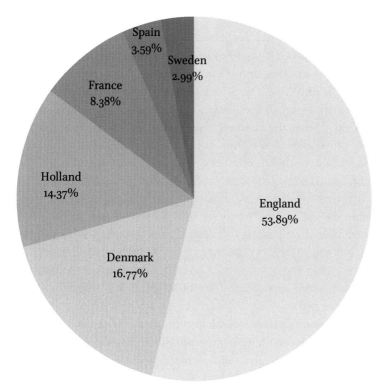

FIGURE 5   Ships loaded with timber in 1760, according to the Count of Aranda's report
SOURCES: AUTHOR'S OWN AFTER AHN, ESTADO, LEG. 4758, EXP. 3. "EL
INFORME DEL CONDE DE ARANDA DE DANZIG".

Warsaw, where the embassy was based, in September, he wrote a report to minister Ricardo Wall, in which he explained several aspects of the harbour's trade, such as the local merchants' eagerness to transact directly with Spain, and estimated how much the harbour's commerce was worth. Aranda also included a section on the export of wood for naval construction in his registers of sea traffic for 1760.[67] According to his observations, 167 ships left the city loaded with timber during the year, of which 90 were bound for Great Britain, 28 for Denmark, 24 for Holland, 14 for France, 5 for Sweden, and 6 for Spain.

These estimates prove that the Crown's decision to increase Spain's commercial presence in the Baltic was the right call, because—as illustrated by timber exports—over half of the region's exports were bound for Great Britain (nearly 54 per cent in 1760), which was Spain's main antagonist for Atlantic

---

67   AHN, Estado, leg. 4758, exp. 3, El informe del Conde de Aranda en Danzig.

supremacy. For this reason, Aranda tried to resume contact with the Schultz family, at the time led by Gabriel Gotlieb, offering the same conditions presented previously by his brother Juan Felipe and Mathy to consul Luis Perrot. Gabriel offered to bring masting from Kronstad (Russia), Riga, and East Prussia to Gdańsk/Danzig and hand them over to Spanish navy officers, so they could arrange their transport to the Iberian Peninsula. He also offered some boards from Poland.[68] Since the contract with Aragorri and Retortillo had just been signed, the proposal was ruled out, but the new *asentistas* still used the Gdańsk/Danzig harbour to find the timber agreed in their contract.[69]

Based on the contract signed in January 1761, the Secretary of the Navy asked the naval departments to estimate their masting and planking needs for 1762. Cartagena requested 334 masts of varying thickness (11½–4½ inches) and length (50–26 cubits); as well as 108 *arbolillos*,[70] 1,000 boards 3 inches thick, 12 inches wide, and 17 cubits long; another 1,000 2½ inches thick, 12 inches wide, and 15 cubits long; and 2,000 regular boards; altogether, accounting for the discount offered by the *asentista*, the cost of this order was 1,255,704 reales de vellón[71] (see Tables 5–8).

In Cádiz – La Carraca, the navy intendent requested 594 masts of varying thickness (12½–4½ inches) and length (50–20 cubits); 240 assorted *arbolillos*; 500 *berlingas*; 2,000 boards 2 inches thick, 12 inches wide, and 18 cubits long; 2,000 2½ inches thick, 12 inches wide, and 18 cubits long; and 2,000 regular boards, for 1,344,588 reales de vellón. Finally, El Ferrol ordered 762 masts of varying thickness (12½–4½ inches) and length (48–25 cubits); 480 assorted *arbolillos*; 800 *berlingas*; 2,000 boards 2 inches thick, 12 inches wide, and 18 cubits long; 2,000 2½ inches thick, 12 inches wide, and 18 cubits long; 10,000 regular boards; and another 8,000 for a private order, for 2,205,846 reales de vellón.[72] In addition to all of this, the minister Julián de Arriaga ordered Retortillo another 63,000 boards of varying thickness, between 1 and 4½ inches thick, for El Ferrol. This order was meant to supply boards for the repair of

---

68   AGS, SMA, Asientos, leg. 616, El Conde de Aranda dice que habiendo estado en Dantzick, le propusieron porción de Arboladuras.

69   AGS, SMA, Asientos, leg. 616, Factura por 5703 tablones de madera para señor Arragori. Danzig, 31 de octubre de 1761.

70   Eng. baulks.

71   AGS, SMA, Asientos, leg. 616, Para poder sacar el importe al. Poco más o menos de la Arboladura que el Asentista debe proveer para este año de 1762.

72   Ibid.

TABLE 5   Estimate of masting needs for the construction of ships-of-the-line in Cartagena (1761)

Number of pieces	Inches thick	Cubits long	Total cubits	Price per piece [r.v.]a
8	11½	50	400	184
22	10½	50	1,100	160
8	11½	50	400	184
22	10½	50	1,100	160
15	10½	38	570	160
8	11	38	304	176
8	9½	48	384	108
30	10½	46	1,380	160
30	9½	41	1,230	108
15	7½	39	585	40
8	7	38	304	34
8	7	48	384	34
4	6	38	152	24
8	5½	30	240	18
8	6	26	208	24
12	4½	26	312	12
30	4	24	720	12
8	5	26	208	14
8	4	28	224	12
4	4	26	104	12
4	3	14	*Arbolillos*	180

a   r.v. = reales de vellón.
SOURCE: AGS, SMA, ASIENTOS, LEG. 616

12 ships of the line over two years.[73] These orders were handed over to Retortillo's representatives in the shipyards. In Cartagena, the *asentista* was represented by Don Joseph Miguel de Ojanguren,[74] and in Cádiz – La Carraca by

73   Ibid., Nota de tablonería de pino del Norte que se podrá necesitar en el departamento de El Ferrol para provisión de dos años; considerando en cada año deberse carenar seis navíos.
74   AGS, Tribunal Mayor de Cuentas (TMC), leg. 4252, Señor don Manuel de la Riva, comisario de provincia de Marina y tesorero de ella en este Departamento del dinero de Real Haci-

TABLE 6   Estimate of masting needs for the construction of frigates in Cartagena (1761)

Number of pieces	Inches thick	Cubits long	Total cubits	Price per piece [r.v.][a]
5	11	47	235	176
5	10	46	230	136
5	8	42	210	56
5	11½	31	155	184
5	10	45	225	136
5	9½	40	200	108
8	6½	30	240	30
5	5	30	150	14
9	6	30	270	24
5	5½	28	140	18
5	4	22	*Arbolillos*	180
7	3½	18	*Arbolillos*	180
9	3	20	*Arbolillos*	180
5	3½	20	*Arbolillos*	180
5	3	20	*Arbolillos*	180
5	2½	22	*Berlingas*[b]	20

a   r.v. = reales de vellón.
b   Eng. spars.
SOURCE: AGS, SMA, ASIENTOS, LEG. 616

Don Alonso Joseph García.[75] These men were responsible for overseeing the ships bringing the timber and also took payment for the deliveries.

In order to meet his commitments, Aragorri and Retortillo, established links with the commercial houses of Thomas and Adrián Hope and company[76] and Horneca and company through the mediation of the Spanish consul in Amsterdam, Don Juan Manuel de Uriondo. These commercial houses worked in the Baltic markets from Szczecin/Stettin to Riga, acting as go-betweens with

      enda del cargo de V.M. pague a Joseph Miguel de Ojanguren apoderado de don Gerónimo Retortillo.

75   AGS, TMC, leg. 4164, Asentista de Arboladura Partida 21.
76   Hope also worked with the French Navy, which it supplied with masts and planks from the southern Baltic and Riga. Robert G. Albion, *Forests and Seapower* (Cambridge, MA: Harvard University Press, 1926), 287–288.

TABLE 7   Estimate of masting needs for the construction of *chambequines* in Cartagena (1761)

Number of pieces	Inches thick	Cubits long	Total cubits	Price per piece [r.v.][a]
2	11	48	96	176
2	10½	46	92	160
2	6	30	60	24
9	6½	46	414	30
5	4	35	175	12
3	7½	44	132	40
3	6	38	114	24
2	3	20	*Arbolillos*	180
2	3½	16	*Arbolillos*	180
2	5½	36	72	18
2	3	16	*Arbolillos*	180

a   r.v. = reales de vellón.
SOURCE: AGS, SMA, ASIENTOS, LEG. 616

TABLE 8   Estimate of masting needs for the construction of *xebecs* in Cartagena (1761)

Number of pieces	Inches thick	Cubits long	Total cubits	Price per piece [r.v.][a]
5	10½	46	230	160
5	10	40	200	136
5	6	30	150	24
22	6	46	1,012	24
12	4	36	432	12
5	7½	45	225	40
5	7	40	200	34
5	3½	20	*Arbolillos*	180
5	3	36	*Arbolillos*	180
5	5½	36	180	18
5	3	20	*Arbolillos*	180

a   r.v. = reales de vellón
SOURCE: AGS, SMA, ASIENTOS, LEG. 616

the naval departments. In 1762, for instance, in Riga they bought 876 masts between 12 and 4/9, and 4 and 2/9 hands thick, and 22,500 boards 1, 2, and 2½ inches thick. In addition, Horneca bought 8,000 pine boards 3, 3½, and 4½ inches thick in Gdańsk/Danzig and Riga; Hope bought 358 masts 12 to 4/9 and 7 to ¹/₉ hands thick. Some of these masts and boards were loaded in Riga in five ships, four for Cádiz and one for El Ferrol. However, in March 1762 the Secretary of the Navy ordered the latter delivery to be halted, as it was at risk of being lost after Spain entered the Seven Years' War in January. Hope sent a letter to Consul Uriondo to have this decision reverted, but to no avail, and the ships did not sail until the following year.[77] In Riga, Hope and Horneca contacted Blankenhagen Oom & Co. to select the best masts from the Duchy of Lithuania, as well as to ship the masting to Amsterdam and Spain. Several vessels were loaded under their supervision between 1761 and 1763, of which the following have been identified: *Santa Cornelia Germina*, captained by Claas Paulsen (see Table 9); *Dama Ildegunda*, whose master was Ide Douwes; *Joven Corneille*, captained by Dirk Hope; *Dama Isabel*, by Juan Knapp; *Elena*, by Theunis Felles; *Dama Cathalina*, by Juan Caan, and *Cordero*, by Telle Vircksz.[78]

A novelty during this period is that masts began to be ordered in advance from local merchants in Riga and Königsberg, and these passed down the orders to their agents in the hinterland. For instance, in 1763, Fridrich Adolf Saturgus, from Königsberg, ordered a load of timber, without specifying its destination, from Józef Jabłoński, commissar of river transport for the tycoon Radziwiłł family. Afterwards, the commissar asked Jan Bachman, forest agent of the Radziwiłłs, for a cargo of masts and bowsprits.[79] Unfortunately, the record does not mention the destination of the wood, but it can be assumed that most contracts adopted similar models, through commercial representatives in the different harbours.

This increased the demand for wood by European naval powers, as Retortillo reported to Secretary Arriaga in a letter dated 7 October 1763:

> I must inform Your Excellency of the issues finding masts 10 to 12.5 hands in Riga, as per the [arsenals'] orders, and it is now impossible to meet this, not because there is a scarcity of masting in that harbour, but because what can be brought next spring cannot suffice for all that is needed by

---

77   AGS, SMA, Asientos, leg. 616, Don Gerónimo de Retortillo, asentista de arboladura para la Provisión en los tres Departamentos de Marina.

78   Ibid., El Corregidor y Regidores de la Imperial Ciudad de Riga.

79   Archiwum Główne Akt Dawnych (AGAD), Warszawskie Archiwum Radziwiłłów (WAR), Kolekcja XX-Handel Rzeczny (1608–1892), exp. 17 y 32.

TABLE 9   Invoice for the boards leaving from Gdańsk/Danzig to El Ferrol aboard the Dutch ship *Santa Cornelia Germina* on 31 October 1761

Boards	Inches, width, and thickness	Length in feet	Florins	60 boards	Total florins
1,000	1 and 12	12	36	60 boards	600
1,400	1.5 and 12	32–34	145	Idem	3,483
700	2 and 12	12–14	90	Idem	1,050
300	2.5 and 12	32–36	300	Idem	1,500
200	3 and 12	32–36	360	Idem	1,200
300	3.5 and 12	32–36	445	Idem	2,225
400	4 and 12	32–36	520	Idem	3,452
					13,510
Expenses and exit duties			2,702		3,188
Danzig 3% commission			486		
					16,698
This amount in Danzig florins, 16,698, 395 *dineros* per pound, amount to Dutch florins					7,608 and 6
Expenses in Amsterdam					
Insurance 4.5 %			342 and 6		
Brokerage 0.5%			38 and 8		532 and 14
Commission 2%			152		8,141
For the ship's hire, 300 last (pol. *Łaszt*) adjusted to			7,500		8,350
Repairs 10%			750		
Sombrero			100		
Dutch florins					16,491
At 8¾ reales de vellón per florin					144,395 reales de vellón

In El Ferrol, the boards measured 2,360 cubic cubits, at 61 reales and 6 maravedís per cubit.

Note: The *asentistas* were paid 34 reales per cubit for the transport from El Ferrol to Cádiz, and it must be taken into account that in peacetime transport was a good deal cheaper.
SOURCE: AGS, SMA, ASIENTOS, LEG. 616

us and other naval powers, especially France, which has just put down a large order [...] with this information [about the demand for masting] you will be able to make your provisions and put down the orders to ensure supply of all that is needed, because if you only count on what is coming down every year to Riga no satisfaction may be expected, because little masting of 10 to 12.5 hands comes, and the other powers also have their needs.[80]

In September 1764, the Minister of the Navy, after consulting the navy intendent in Cádiz, Don Juan Gerbaut, gave the order to reserve 258 thick, top-quality masts, to be brought to Riga in the following spring. The contracts involved five landholders in the Polish-Lithuanian Commonwealth (see Table 10), who committed to supply masts between 9¼ and 14⅕ hands thick.[81] Retortillo also asked Blankenhagen Oom for a report on all the masts that had been brought to Riga during 1764. The report was surprising because it was found that the 21 forest holders in the Kingdom of Poland and the Duchy of Lithuania had delivered only 696 masts to the harbour.[82] This clearly shows that the sector had not yet recovered from the interruption in trade brought about by the Seven Years'

TABLE 10   Landholders from the Polish-Lithuanian Commonwealth with whom contracts were signed to supply masting to Cádiz in 1765

Name	Number of masts	Hands thick in Spain
Reuhutz	94	9¾–13¼
Radziwiłł	60	9¼–12
Massalski	30	9¾–11½
Swadkowski	23	9¾–14⅕
Kazimirski	51	9¾–11½

SOURCE: AGS, SMA, ASIENTOS, LEG. 616

---

80   debo hacer presente a VE que sabiendo por experiencia que con dificultad se encuentran en Riga los palos de 10 a 12,5 palmos que son los que piden [los arsenales], lo es por ahora imposible tomar empeño alguno de esta naturaleza no porque falta arboladura de venta en aquel puerto sino que la que ha de bajar en la próxima primavera no puede ser suficiente para el surtimiento que necesitan como nosotros, las demás potencias marítimas y particularmente la Francia que acaba de hacer una contrata para una porción considerable. [...] con esta noticia [sobre las necesidades de arboladura] y anticipación podrá el suplicante hacer sus cálculos y dar desde ahora sus órdenes para asegurarse con los propietarios de los montes, de todas las cantidades que sean necesarias pues si solo cuenta con lo que baja anualmente a Riga no se puede esperar ningún desempeño tanto por que vienen muy pocas perchas de 10 a 12,5 palmos como porque las demás potencias tienen igualmente sus urgencias. (AGS, SMA, Asientos, leg. 616, Arboladura del Norte).

81   Ibid., Facturas de Arboladura del Norte.

82   Ibid., Resumen de la Arboladura que ha bajado a Riga en la primavera de 1764.

War, because, before the conflict, Riga received between 1,800 and 3,800 masts per year from its hinterland.[83]

Another problem concerning wood supply, especially for countries that did not have major commercial fleets in the North—such as France, Portugal, and Spain—was the scarcity and loss of shipping. According to a report filed on 16 September 1766 by Juan Gerbaut—who was preparing a new *asiento* for the following year—50 ships had been lost in Baltic and northern waters in 1761 alone. Five years later, shipowners, especially the Dutch, were still reeling from these loses, because freighting left little margin for profit. For this reason, Gerbaut was in favour of making orders for pine masting in advance.[84]

For this reason, advance orders and payments for thick masts of different sizes were made to Polish-Lithuanian forest owners throughout 1766 and 1767. In 1766, 273 masts of varying sizes, from 12½ to 9¼ hands thick and from 46 to 37½ cubits long, were ordered from ten forest holders (see Table 11).[85] In 1767,

TABLE 11 Landholders from the Polish-Lithuanian Commonwealth with whom contracts were signed to supply masting to Cádiz in 1766

Name	Number of masts	Hands thick in Spain
Reuhutz	81	9⅙–11¼
Radziwiłł	56	9⅙–12
Massalski B.	12	9⅙–10½
Massalski K.	12	10⅐–11
Swadkowski	9	9⅙–12
Jagucianski	11	9⅙–11
Wyszynski	30	9⅙–12
Dernattowitz	41	9⅙–12
Oskieko	10	10⅕–12
Piaskowski	11	10⅐–11⅓

SOURCE: AGS, SMA, ASIENTOS, LEG. 616

83 Darius Žiemelis, "The Structure and Scope of the Foreign Trade of the Polish-Lithuanian Commonwealth in the 16th to 18th Centuries: The Case of the Grand Duchy of Lithuania", *Lithuanian Historical Studies* 17 (2012), 114; Pierrick Pourchasse, "The Control of Maritime Traffic and Exported Products in the Baltic Area in Early Modern Times: Eighteenth-Century Riga", *Revue Historique* 686, no. 2 (2018), 389–390.

84 AGS, SMA, Asientos, leg. 616, Condiciones que se enviaron para el Asiento.

85 Ibid., Resumen de las 10 partidas de Arboladura gruesa que han de bajar al puerto de Riga en la primavera de 1766.

the orders amounted to 503 pine masts (9½ to 12⅓ hands thick and 49 and 40 cubits long) from 15 landholders (see Table 12).[86]

The growing number of contractors is to be noticed; while orders were put to only five landholders in 1765, the figure had risen to 15 two years later. This can be interpreted in two ways: first, the wish to diversify suppliers, so that if one failed to meet their order this would have a limited impact on the overall supply of masting; second, that wood was beginning to run short in some areas, which seems to be supported by the fact that only two contractors feature in all three contracts (Radziwiłł and Massalski).[87] This seems to have forced Spain to

TABLE 12   Landholders from the Polish-Lithuanian Commonwealth with whom contracts were signed to supply masting to Cádiz in 1767

Name	Number of masts	Hands thick in Spain
Radziwiłł	60	
Massalski	15	
Wyszynski	43	
Czartoryski	108	
Czarniawski	17	
Darelowicz	40	*Not broken down by*
Drobiszewski	12	*contractor. The contracts*
Wiskinski	27	*only specified a range*
Malinowski 1	30	*between 9½ and 12⅓.*
Gruszecki	21	
Malinowski 2	15	
Iskieike	20	
Joseph Soltan	8	
Czerwinski	15	
Ukrainsche	72	

SOURCE: AGS, SMA, ASIENTOS, LEG. 616

86   Ibid., Condiciones a que don Gerónimo de Retortillo se obliga a conducir a Cádiz la porción de Arboladura que ha de bajar a Riga en la primavera del año próximo de 1767.

87   Unfortunately, contracts do not include first names, which would help to identify exactly the sources of wood. However, there is little doubt that the names Radziwiłł and Massalski refer to two of the most powerful families in the north-east and east of the Polish-Lithuanian Commonwealth. They owned enormous forests in the territories of modern Poland, Lithuania, Belarus, and Russia. Zbigniew Anusik and Andrzej Stroynowski, "Radziwiłłowie w epoce saskiej. Zarys dziejów politycznych i majątkowych", *Acta Universitatis*

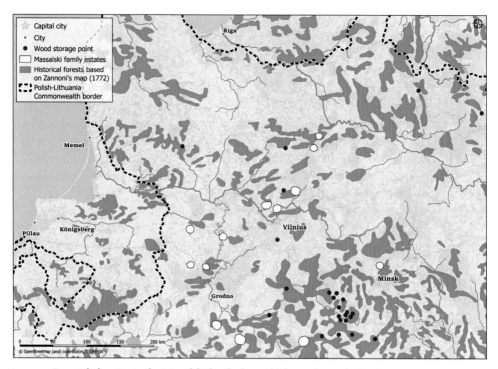

MAP 4   Forests belonging to the Massalski family, from which wood was obtained.
SOURCE: LOOK4GIS-LUKASZ BRYLAK BASED ON QGIS SOFTWARE

open the range of suppliers. Map 4 shows the lands that were under the control of the Massalski magnate family.

Aragorri's *asiento* expired on 31 December 1766, and in November the Marquis of Muzquiz, secretary of finances and war, ordered the naval departments and the harbours of Bilbao and San Sebastián to publish conditions for a new *asiento* for northern masting, giving 40 days to present offers. On 7 January 1767, the minister reported that no tenders had been submitted, probably as a result of the Crown's strict price – quality conditions, as well as the constant difficulties with transport, largely left in Dutch hands through the Hope and Horneca houses.[88] In fact, these merchants emphasized the obstacles that hampered the transport of timber to Spain, first because of the Seven Years' War and the activity of British ships, and second because of a conflict with Amsterdam dock workers.[89]

Lodziensis 33 (1989), 29–58; Ewa Dubas-Urwanowicz y Jerzy Urwanowicza (eds.), *Wobec króla i Rzeczpospolitej. Magnateria w XVI–XVIII wieku* (Krakow: Avalon, 2012).

88   AGS, SMA, Asientos, leg. 616, Con motivo de haberse publicado en los tres departamentos y en Bilbao y San Sebastián.

89   Ibid., Carta escrita a don Simón de Aragorri por los señores Hope y compañía.

As such, in the period 1767–1771, the supply of timber was organized through *ad hoc* contracts, most of which were executed by Aragorri and his representative Retortillo, for instance in 1767, when he committed to deliver 503 masts, 20,000 boards 1 inch thick, 3,000 1.5 inches thick, 3,000 2.5 inches thick, and 3,000 2 inches thick, at 58 reales the cubic cubit.[90] According to the correspondence between the El Ferrol and Cádiz intendents, Pedro de Hordeñana and Juan Gerbaut, respectively, and the minister Julián de Arriaga, both officials were in favour of accepting Aragorri's tender. Meanwhile, the contractor's representative in Gdańsk/Danzig was informed that 14,000 boards had already been put aside in this harbour to be shipped during the summer of 1767, information that he immediately passed on to the Secretary of the Navy.[91]

The expense records of the navy's treasury in the department of Cartagena from 1 January 1769 to 31 December 1771 illustrate maritime traffic between the Baltic and Spain.[92] All the cargoes of northern timber were managed by the Spanish consul in Amsterdam, showing that the system was entirely reliant on Dutch shipping, organized by Hope and Horneca. In these three years, Don Juan de Zalvide, navy treasurer in Cartagena, paid for nine cargoes of northern masts and planks, delivered by the Dutch fluyt *Brocker* (twice), *Las Dos Hermanas* (twice), *San Pedro* (twice), and *Gesina Petronela*, and by the Danish *Oriente* (twice). The harbour of origin is only specified in two instances, that of the *Oriente*, captained by Olaf Engelbrecht Petersen (Gdańsk/Danzig),[93] and the *Gesina Petronela*, captained by Jacobo Jonker (Riga).[94] No departure harbour is mentioned in the records for the other deliveries, but in all cases, they called in Amsterdam, where they were inspected by Uriondo.

Cádiz and El Ferrol were busier than Cartagena, but this still fell short of the arsenals' demands, and two additional offers to supply northern masting and planking were submitted as a result. The first concerns a contract for 263 large masts from Russia, presented by the Rey and Brandenburg commercial house, led by the Lyon-born Guillermo Rey and Juan Federico Brandenburg,[95] from Fredrikshald, Sweden. The company offered to request permission from Empress Catherine II "to fell masts in her forests to the exact required

90   Ibid., Don Gerónimo de Retortillo propone traer los 503 palos.
91   Ibid., El Comisionado que tengo en Danzig en la inteligencia.
92   AGS, TMC, leg. 4257, Diferentes patrones de embarcaciones por fletes de cáñamo, madera y otros géneros desde 1º de enero de 1769 hasta 31 de diciembre de 1771.
93   Ibid., Libranza del 17 de febrero de 1770.
94   Ibid., Libranza del 3 de octubre de 1770.
95   He was also Russian consul in Cádiz. AHN, Estado, leg. 5322, exp. 75.

measurements".[96] The deal did not come to fruition, but the consul continued exploring possibilities to work with the Spanish Secretary of the Navy, which finally crystallized in the 1770s, when he worked together with Felipe Chone, from Bilbao, who was awarded the northern timber *asiento* in 1772.[97] The second offer was submitted by the Cádiz-based company Herman & Ellermann & Schlieper, which had been delivering flax, and small quantities of "northern pines and planks", from Hamburg to Cádiz since the 1750s.[98] In this instance, the German firm offered to bring 15,000 boards of various thickness to Cádiz, but the offer was rejected, although in contrast with Rey and Brandenburg, Herman & Ellermann & Schlieper became associated with Chone's *asiento*, as shown by the records of La Carraca for October 1776, when the navy treasurer in this department paid the company's representatives 170,727 reales de vellón for the transport of 1,448 cubic cubits, comprising 207 pine mother beams and seven *tabloncillos*, and 105,553 reales for 12,418 boards at 8.5 reales de vellón each.[99] Herman & Ellermann & Schlieper began working for the navy on a larger scale in the 1780s, when the northern *asientos* came under the management of the Banco de San Carlos.[100]

5   The French Naval Construction System and Soto's, Chone's, and Marraci's *Asientos* (1772–1782)

As noted, sea traffic was light in 1762 because of the British blockade. When the war ended with the Treaty of Paris on 10 February 1763, the Spanish navy entered a new stage in its modernization process, after the war had demonstrated its inability to protect commercial routes and defend the American colonies against the disciplined and well-organized Royal Navy. The Secretary

---

96   "para mandar cortar arboladuras solas en sus montes de las medidas exactas y las que se pidiese". AGS, SMA, Asientos, leg. 616, Palos de arboladura que ofrece proveer la casa de comercio Rey y Brandemburg.

97   María Guadalupe Carrasco González, "Cádiz y el Báltico. Casas comerciales suecas en Cádiz (1780–1800)", in Alberto Ramos Santana (ed.), *Comercio y Navegación entre España y Suecia (siglos X–XX)* (Cádiz: Universidad de Cádiz, 2000), 338–343; Torres Sánchez, *Military Entrepreneurs*, 202–203.

98   AGS, TMC, leg. 4164, Madera año de 1764, partida sexta.

99   Ibid., Señor don Miguel de Bertodano, comisario de Provincia y tesorero de Marina.

100   Juan Torrejón Chaves, "La madera báltica, Suecia y España (siglo XVIII)", in Ramos Santana, *Comercio y navegación entre España y Suecia*, 182; Ángela Fernández Cañas-Baliña, "Tráfico comercial con el Báltico y el Mar del Norte: suecos y daneses, su relación con Cádiz a través del Diario Marítimo de la Vigía, y el Registro de Sund (1789–1800)", *Baetica* 41 (2021), 257–258.

of State, Pablo Jerónimo Grimaldi, who was acquainted with the operation of French arsenals—he had been ambassador in Versailles (1761–1763)—prompted Charles III to ask the French navy minister, Étienne-François de Choiseul, Duke of Choiseul, for a French naval constructor on loan. The initiative aimed to tighten the military links between the two Bourbon crowns, which began reorganizing their armies and navies in the immediate aftermath of their defeat in the Seven Years' War. The Duke of Choiseul negotiated this loan with Grimaldi, and agreed to lend Spain a young and ambitious shipbuilder, Francisco Gautier. When Gautier arrived in Madrid in early 1765, he was introduced to the king, who was presented with his designs for a 70-gun ship-of-the-line and a 26-gun frigate. The engineer was soon deployed to the shipyard of Guarnizo, but he first visited the mounts of Cantabria to assess the oak forests there. Then the shipyard decided to finish the six 70-gun ships and four frigates corresponding to Manuel de Zubiría's *asiento*, which had not made any progress in months. Although the wood was cut according to the former British construction system adopted by Jorge Juan and British constructors Richard Rooth, David Howell, Edward Bryant, Matthew Mullan, and Almond Hill in the 1750s, Gautier introduced modifications that, while increasing the cost, managed to get the ships-of-the-line launched between October 1765 and February 1769, and the frigates between July 1767 and August 1768. His skill and hard work earned him the appointment as ship construction and repair director on 25 April 1769, working autonomously from naval department commanders and intendents, and as army colonel engineer. This appointment was a turning point for Spanish naval construction, which thenceforth adopted the French construction model under Gautier's supervision. This model remained the norm until 1782, when Gautier resigned from his post and returned to France, where he was appointed director of construction in Toulon.[101]

This change affected the south Baltic timber *asientos*. As noted, between 1767 and 1771 the system operated through individual contracts, which were, at any rate, implemented by Aragorri and Retortillo, which shows that the Spanish Crown was determined to keep the business in Spanish hands. This policy

---

101   José Patricio Merino Navarro, *La Armada Española en el siglo XVIII* (Madrid: Fundación Universitaria Española, 1981), 123–127; José Ignacio González-Aller Hierro, *Modelos de arsenal del Museo Naval evolución de la construcción naval española, siglos XVII–XVIII* (Barcelona: Lunwerg, 2004), 106–119 and 190–193; José María Blanco Núñez, *La Armada Española en la segunda mitad del siglo XVIII* (Barcelona: IZAR Construcciones Navales, 2004), 183–189; Valdez-Bubnov, *Poder naval y modernización del Estado*, 322–339; José María Sánchez Carrión, *De constructores a ingenieros de marina. Salto tecnológico y profesional impulsado por Francisco Gautier* (Madrid: Fondo Editorial de Ingeniería Naval, 2013), 411–412.

was maintained with the return of the *asiento* system, with the award of a long-term contract to Felipe Chone, representative of the Soto, in 1772.[102] This *asiento* specified measures for masting, *arbolillos, berlingas,* and planking to be delivered to the naval departments. The interesting thing about this contract was that Riga, or alternatively Saint Petersburg, was specified as the timber's port of departure. The Crown emphasized that:

> the Masting must be [...] freshly cut, without too much sap or irregular knots that weaken the masts; they must be straight and free from bent veins, to the satisfaction of the inspecting officials; similarly, boards must be of the best quality, and especially those one and a half inches thick or less, which are used for many purposes.[103]

The economic committee of naval departments, made up of naval officials, which met regularly, convened to discuss the changeover to Gautier's system in September 1772. The committee stressed that wood *asentistas* were to be especially careful to keep the dimensions of the timber pieces delivered to Gautier's specifications.[104]

Despite Gautier's detailed instructions, Chone resumed the practice of asking individual arsenals for information about the parts that they needed, as was done back in the 1760s. This information was sent to Chone's agents in Riga and Saint Petersburg, who put down advance orders to local merchants and landowners. The inclusion of Saint Petersburg in the contract is a reflection of Spain's foreign policy in the 1770s and 1780s, when Charles III tried to establish closer links with Catherine II, which resulted in increased commercial relations with Rey and Brandenburg in both Russian and Spanish harbours.[105] The Spanish presence in Russia was also boosted by Antonio Colombí

102   Óscar Riezu Elizalde y Rafael Torres Sánchez, "¿En qué consistió el triunfo del Estado Forestal? Contractor State y los asentistas de madera del siglo XVIII", *Studia Historica. Historia Moderna* 43, no. 1 (2021), 217.

103   "la Arboladura ha de ser [...] del corte más reciente, sin mucho sámago, ni de nudos irregulares que debilitan las Perchas; derechas éstas, sin veta demasiado torcida, y a satisfacción de los Oficiales comisionados para reconocerlas, y recibirlas; igualmente la Tablazón deberá ser de buena calidad y especialmente las clases de pulgada y media abajo, que se emplean en obras blancas". AGS, SMA, Asientos, leg. 788, Asiento general de Provisión de Arboladura y Tablonería de Pino del Norte a cargo de don Felipe Chone por termino de seis años.

104   Archivo Naval de Cartagena (ANC), Junta Económica del Departamento (JED), libros de acuerdos vol. 2729, tomo 1 (1772).

105   Carrasco González, "Cádiz y el Báltico", 338–343; Cezary Taracha, "Algunas consideraciones sobre la cuestión rusa y turca en la política española de la época de Carlos

y Payet, who set up a commercial house in Saint Petersburg in 1773. First, he acted on behalf of the Milans house, but he took over the business in 1776, creating a permanent base in Saint Petersburg. His commercial success and his connections with Russian merchants and politicians earned him the appointment as Spanish consul in Russia. In 1786, he was promoted to general consul, in which position he served the interests of the Spanish Crown until his death in 1811.[106] Another important figure for Spanish – Russian commercial relations was Pedro Normande, who, like Colombí, participated actively in the transport of timber and other raw materials for the Spanish navy. His commercial skills and command of the Russian language were behind his appointment as temporary Spanish minister in Russia between 1783 and 1788, when the Count of Floridablanca appointed a new ambassador, Miguel de Gálvez y Gallardo.[107]

A novel element was the voyages to the Baltic of *asentistas* in search of new contracts, such as Felipe Chone's in 1768. Previously, Chone had supplied the Spanish navy with Baltic hemp. On this trip, he visited Viborg, Frederishaven, Saint Petersburg, and other harbours in the Bay of Finland, where he examined masting and planking. His measurements revealed differences in size with regard to the masts being sent from Riga to the Dutch and British markets. In addition, in Wiborg, Chone saw three ships with larger masts than those sold in Riga, which confirms that Riga acted as an *entrepôt* for products sourced in the Bay of Finland and the harbour's hinterland in the Polish-Lithuanian Commonwealth.[108]

When Soto and Chone's *asiento* was signed, the Spanish consul in Amsterdam, Juan Manuel de Uriondo, created a new network of informants in two key harbours, Gdańsk/Danzig and Riga. In addition, as noted, relations with Russia became closer from 1773, with the arrival of the Catalonian house Milans and their agent Colombí in Saint Petersburg. Meanwhile, Uriondo's agents, led by

---

111", *Teka Komisji Historycznej* 9 (2012), 53–75; Óscar Recio Morales, "Los militares de la Ilustración y la construcción del Este de Europa en España", *Itinerarios. Revista de Estudios Lingüísticos, Literarios, Históricos y Antropológicos* 31 (2020), 34–56; Fernández Cañas-Baliña, "Tráfico comercial con el Báltico y el Mar del Norte", 231–266.

106   Pradells-Nadal, "Los cónsules españoles del siglo XVIII", 217, 224; Agustín Guimerá Ravina, "La casa Milans una empresa catalana en Rusia (1773–1779)", *Pedralbes: Revista d'Historia Moderna* 18, no. 1 (1998), 83–92; Rafael Torres Sánchez, "Contractor State and Mercantilism. The Spanish-Navy Hemp, Rigging and Sailcloth Supply Policy in the Second Half of the Eighteenth Century", in Richard Harding and Sergio Solbes Ferri (eds.), *The Contractor State and Its Implications, 1659–1815* (Las Palmas: Universidad de Las Palmas de Gran Canaria, 2012), 332.

107   Recio Morales, "Los militares de la Ilustración", 48–50.

108   AGS, Secretaría y Superintendencia de Hacienda (SSH), leg. 55, Los adjuntos papeles de don Felipe Chone.

Antonio de Cuyper, established links with Dutch merchants who organized the purchase and dispatch of the planking that arrived in Gdańsk/Danzig down the Vistula River. In Riga, the Spanish consul was represented by Herman Fromhold, who was entrusted with finding masting and planking for the Spanish navy.[109] The connection with Gdańsk/Danzig-based merchants clearly shows that, although the *asiento* officially limited purchases to the harbours of Riga and, alternatively, Saint Petersburg, this condition could not be met, because timber was scarce. This is also confirmed in a letter from Fromhold, who responded to the consul's order of 129 large masts for El Ferrol by claiming that "the English [sic] buy there [in Riga] almost all the masting in stock and what is expected to come down from the Polish hills in the spring on top of that".[110]

The Royal Navy's growing purchases in the Baltic markets were triggered by the beginning of the American War of Independence in 1775. The conflict severed the oak and pine supply lines from New England, so the British were forced to turn to the Baltic to avoid running short of materials for naval construction.[111] The British demand prevented Saint Petersburg and Riga from meeting orders by other European navies, and Chone had to reach out to other southern Baltic harbours instead, as recorded in the delivery registries of El Ferrol for 1776: the Dutch fluyt *Ana y Samuel*, captained by Hilke Sibrands, arrived in the Galician port with 4,979 pine boards, 60 beams, and 142 *cuartones*;[112] the Dutch ship *Astrea*, captained by Martin Zadach, arrived from Gdańsk/Danzig with 6,609 boards of various sizes;[113] the 400-ton pink *Suartelter*, captained by Pedro Peterson, arrived from Gdańsk/Danzig loaded with planks;[114] and the Dutch schooner *Rosa de Oro*, captained by Willem Vandervind, arrived from Memel with 3,985 boards of different sizes.[115] Other records from El Ferrol, La Carraca,

109 AGS, SSH, leg. 55, He reconocido la carta de don Juan Manuel de Urionda.

110 "los Ingleses hacen allí [en Riga] contratas por casi toda arboladura existente y la que ha de bajar de los montes de Polonia en la Primavera". Ibid., Según los avisos que de cuando en cuando me vienen de don Herman Fromhold.

111 Albion, *Forests and Seapower*, 144–145; Paul Walden Bamford, "France and the American Market in Naval Timber and Masts, 1776–1786", *Journal of Economic History* 12, no. 1 (1952), 21–22; R.J.B. Knight, "New England Forests and British Seapower: Albion Revised", *American Neptune* 46 (1986), 223–224; James Davey, *The Transformation of British Naval Strategy: Seapower and Supply in Northern Europe, 1808–1812* (Suffolk: Boydell, 2012), 26–27.

112 AGS, SMA, Arsenales leg. 635, El día 16 del corriente entró en este puerto la urca holandesa nombrada *Ana y Samuel*.

113 Ibid., El día 13 del corriente dio fondo en este puerto procedente de Danzig.

114 Ibid., En el día de ayer a las cinco de la tarde arribó a este puerto el pingue dantzigues.

115 Ibid., El día 10 del corriente mes entraron y dieron fondo en este puerto procedentes de Memel y Danzig.

and Cartagena prove that the obligation to source timber only from Riga and Saint Petersburg was little more than wishful thinking on the part of the navy, which was under the impression that the timber purchased in these Russian harbours was better. The situation, however, forced naval officials to accept timber from other Baltic regions, which was more accessible but of lesser quality. Purchases were made in cities such as Memel, Pillau, Königsberg, Gdańsk/ Danzig, and Szczecin/Stettin, helping Spanish naval construction to modernize and pick up speed between 1770 and 1790 until Spain became the second naval power globally in the 1780s.

Chone's *asiento* expired in December 1777, and the following contractor to receive the approval of the Secretary of the Navy was Don Carlos María Marraci,[116] whom it should be noted had taken part in Chone's operation as his agent, first in Cartagena and later in Cádiz and El Ferrol.[117] Baltazar Castellini had, for his part, worked with the navy as Marraci's representative from 1763.[118] These precedents allowed Marraci to be awarded a four-year (1778–1782) *asiento* in his own right. He committed to bring first-rate, freshly cut, knot-free, straight timber from Riga and Saint Petersburg,[119] but, as the records from El Ferrol and La Carraca attest, Marraci could not meet this condition, owing to the demand posed by other naval powers, especially Britain.

Therefore, as Chone before him, he had to buy timber from other harbours in the region, as shown in Table 13. In this case, an analysis of maritime traffic from 1779 clearly shows that the main port from where timber was purchased for the Spanish Navy was Gdańsk/Danzig (24 vessels—86 per cent of total movement) and only Memel (two vessels) Riga (one vessel), and Pillau (one vessel) are sporadically mentioned.

During the American War of Independence, in which Spain actively participated from 1779, ships loaded with naval supplies were a prime target, as expressed by Marquis González de Castejón in a letter to the intendents of the naval departments in the Iberian Peninsula in February 1781: "in the middle of the tribulations of war, several ships sent with hemp and northern pine

---

116   Torres Sánchez, *Military Entrepreneurs*, 205–207.

117   Two more offers were made, but these were rejected. One was submitted by Juan Josef Vicente de Michelena, a San Sebastián-based merchant, and the second by Claudio Rochefort, representative of Don Francisco de Paula Henríquez, from Cádiz. AGS, SMA, Arsenales leg. 635, Juan Josef Vicente de Michelena, vecino y del comercio de San Sebastián and Don Claudio Rochefort vecino de Madrid, como apoderado de don Francisco de Paula Henríquez, que lo es de Cádiz.

118   Castellini also represented Felipe Chone in his dealings with the naval department of Cartagena. ANC, JED, libros de acuerdos vol. 2729, tomo 2 (1774).

119   AGS, SMA, Arsenales leg. 635, Don Carlos María Marraci y Compañía, vecinos y del Comercio de esta Corte.

TABLE 13   Ships with timber for naval construction sourced from the Baltic by *asentista* Carlos María Marraci (1779)

Registry date	Ship name	Captain	Port of origin	Port of arrival	Load (pine wood)
20 January 1779	*Fama* Dutch fluyt	Jongeboer	Gdańsk/ Danzig	El Ferrol	2,594 boards and 61 *tozas*
20 January 1779	*Spiering Hokke* Dutch fluyt	Dekker	Gdańsk/ Danzig	El Ferrol	3,203 boards and 37 *tozas*
20 January 1779	*Unión* Dutch fluyt	Kirsch	Gdańsk/ Danzig	El Ferrol	2,851 boards and 87 *tozas*
27 January 1779	*Margarita Susana* Dutch fluyt	Herman Bakker	Riga	El Ferrol	115 pieces of masting and 900 boards
27 January 1779	*Habran y Jacob* Dutch fluyt	Jube Pieters	Memel	El Ferrol	12 pieces of masting and 4,327 boards
27 January 1779	*Señora Agata Francisca* Dutch fluyt	Juan Meyers	Gdańsk/ Danzig	El Ferrol	2,059 boards and 59 *tozas*
7 April 1779	*Floreciente* Dutch ketch	David Hilles	Gdańsk/ Danzig	El Ferrol	3,928 boards and 110 *tozas*
7 April 1779	*Dos Hermanos* Dutch ketch	Emanuel Gotlieb Martens	Gdańsk/ Danzig	El Ferrol	1,851 boards and 41 *tozas*
7 April 1779	*Cristina Magdalena* Dutch ketch	David Jacobs	Gdańsk/ Danzig	El Ferrol	3,495 boards and 60 *tozas*
14 April 1779	*Waaksamey* Dutch fluyt	Natamiel	Gdańsk/ Danzig	El Ferrol	6,679 boards and 38 *tozas*
14 April 1779	*Mercurio Volante* Dutch fluyt	Juan Runchen	Gdańsk/ Danzig	El Ferrol	5,597 first-rate parts and 64 last-rate parts
14 April 1779	*Dama Heta* Dutch fluyt	Hene Jansz	Gdańsk/ Danzig	El Ferrol	1,790 boards and 42 *tozas*
14 May 1779	*Señora Deliana* Dutch ketch	Jechle Spiur	Memel	La Carraca	2,664 boards
15 May 1779	*Estrella de la Mañana* fluyt	Frederico Kinder	Gdańsk/ Danzig	El Ferrol	3,749 boards and 38 *tozas*

TABLE 13   Ships with timber for naval construction sourced from the Baltic (*Cont.*)

Registry date	Ship name	Captain	Port of origin	Port of arrival	Load (pine wood)
29 May 1779	*Vigilancia* fluyt	Jiede Christians	Gdańsk/ Danzig	El Ferrol	2,986 boards and 66 *tozas*
9 June 1779	*Hendrik* Dutch fluyt	Jacobo Cornelis	Gdańsk/ Danzig	El Ferrol	3,897 boards
9 June 1779	*Henrrieta Constancia* Dutch fluyt	Federico Rosenau	Gdańsk/ Danzig	El Ferrol	3,502 boards and 40 *tozas*
26 June 1779	*Navegación* Dutch ketch	Imke Hessels Dries	Gdańsk/ Danzig	El Ferrol	Boards and *tozas*
30 June 1779	*Cevaart* Dutch ketch	-	Gdańsk/ Danzig	El Ferrol	2,158 boards and 54 *tozas*
23 July 1779	*Mercurio* Swedish frigate	Jacobo Frederik	Pillau	La Carraca	1,011 large boards and 3,883 boards
3 August 1779	*Gerardo y Enrique* Dutch vessel	Hilbrand Kersyes Roos	Gdańsk/ Danzig	La Carraca	826 large boards and 6,128 boards
17 August 1779	*Amistad* Dutch vessel	Pedro Jacobs	Gdańsk/ Danzig	La Carraca	595 large boards and 3,437 boards
27 August 1779	*Cornelio y Rodrigo* Dutch fluyt	Pedro Pedros	Gdańsk/ Danzig	La Carraca	1,801 large boards and 4,783 boards
27 August 1779	*Catalina* Dutch fluyt	Doueves Ideiz	Gdańsk/ Danzig	La Carraca	1,346 large boards and 7,225 boards
16 October 1779	*Jonge Jower* Dutch ketch	Cornelis Johanes	Gdańsk/ Danzig	El Ferrol	1,924 large boards and 12 *tozas*
13 November 1779	*Jonge Tromp* Dutch ketch	Pier Hanses	Gdańsk/ Danzig	El Ferrol	1,978 boards
20 November 1779	*Jonge Baitie* Dutch ketch	Arent Aldert Smit	Gdańsk/ Danzig	El Ferrol	2,154 boards
24 November 1779	*Concordia* Dutch fluyt	Teerkes Repko	Gdańsk/ Danzig	El Ferrol	7,881 boards

asentistas, on their way to the departments to supply the King's navy, have been captured".[120]

In order to avoid capture by the Royal Navy's ships patrolling the Channel,[121] contractors such as Pedro Normande and Antonio Colombí began hiring Prussian and Russian ships, which between 1779 and 1783 remained neutral. However, contacts with freighters from these nations was through Cádiz-based commercial houses, such as Marraci's and Colombí's, but also French ones, such as Cayla & Solier & Cabanes & Jugla's,[122] set up in 1766; for their part, Galibert Cayla and Jean Solier[123] established links with Cádiz trade networks as early as 1730. This French commercial house organized at least three deliveries of northern masts and planks in 1783. First, it used its contacts with Pedro Normande to arrange three trips, one from Saint Petersburg, commanded by the Prussian captain Christian Schmidt,[124] and then two more from the same harbour, aboard the Russian ship *Riga*, captained by Federico Reinders, and the Prussian ketch *Las Dos Nietas*, captained by Sietze Paul Schoul.[125] The register from the El Ferrol department for March 1782, signed by Juan Joseph de Samacoyz, records the arrival of several Baltic ships in this port, and masting and planking sent by Antonio Colombí's house in 1781.[126] Interestingly, all this timber was brought aboard Prussian and Russian ships (see Table 14).

These deliveries show that, from the 1780s, the Dutch monopoly over freight operations found new competitors in the ships from neutral nations; the example includes Prussian and Russian ships, but the Swedes and Danish were also arriving more and more frequently.[127]

---

120   "cuando en medio de las dificultades y riesgos de la guerra, cuando en medio de las varias embarcaciones de los asentistas de cáñamo y pino del Norte, que han sido interceptadas por los enemigos se arriesgan aquellos [asentistas] a conducir a los departamentos lo que tanto se necesita para los aprestos de la armada del Rey". AGS, SHH, leg. 51, Avisos y consignaciones de Marina del año 1781.

121   David Syrett, *The Royal Navy in European Waters During the American Revolutionary War* (Columbia: University of South Carolina Press, 1998); Thomas Chávez, *Spain and the Independence of the United States: An Intrinsic Gift* (Albuquerque: University of New Mexico Press, 2002).

122   In the 1760s this company worked with the Dutch houses of Hope and Horneca. Amsterdam City Archives (ACA), archive 5075, notarial records 11452A and 11457, register numbers 428883 and 397528 (1766), https://archief.amsterdam/inventarissen/details/5075/path/348.3.351.

123   Xabier Lamikiz, *Trade and Trust in the Eighteenth-Century Atlantic World: Spanish Merchants and their Overseas Networks* (Suffolk: Boydell Press, 2013), 158–160.

124   AGS, SHH, leg. 51, Avisos y consignaciones de Marina del año 1783.

125   Ibid., Avisos y consignaciones de Marina del año 1784.

126   AGS, SMA, Asientos leg. 368, Consecuente a lo que V.E. se sirvió prevenirme en orden de 28 de diciembre del año próximo pasado.

127   Fernández Cañas-Baliña, "Tráfico comercial con el Báltico y el Mar del Norte", 231–266.

TABLE 14   Report of ships arriving at El Ferrol from the Baltic in Prussian and Russian ships in 1781

Ship name	Nationality and type	Captain's name	Cargo
Beurzs van Emdem	Prussian brigantine	Soen Roclofsz	51 pine masts, 200 *berlingas*, 1,418 boards
Jonge Sybrand	Prussian ketch	Christian Kolm	40 pine masts, 260 *berlingas*, 891 boards
Juffer Margareta	Prussian ketch	Gerbrand Pieters	50 pine masts, 200 *berlingas*, 1,324 boards
Neptuno	Prussian ketch	Barend Dirkez	47 pine masts, 248 *arbolillos*, 1,203 boards
Cobemzel	Imperial galley	Roeloff Griffen	53 pine masts, 600 *berlingas*, 1,162 boards
Beurs van Riga	Russian brigantine	Christian Alzersen	64 pine masts, 200 *berlingas*, 856 boards
Davidof	Russian vessel	Johann Chrisopffer Bobst	Canvas
Lekoy	Russian war frigate	—	Hemp
Micaela	Russian war frigate	—	Hemp

SOURCE: AGS, SMA, ASIENTOS LEG. 368

The Soto – Chone *asiento* plays a central role in the history of the supply of masting, as it inaugurated the peak period in the trade of Baltic timber to Spain, from 1775 to 1790 (see Tables 15 and 16). In these years, as many as 377 Spain-bound ships loaded with timber crossed the Sund;[128] 1776, 1777, and 1778 were the most active years,[129] with 32, 34, and 36 ships, respectively.[130] The 1770s also witnessed the construction of several warships, in whose construction Baltic timber played a significant role, especially in terms of planking and masting, which often amounted to over 35% and 40% of all the wood used in

---

128   Reichert, "El comercio directo de maderas", 150–152.
129   Harbours of Lübeck, Wismar, Wolgast, Rostock, Swinemünde, Szczecin/Stettin, Kolberg, Stolp, Gdańsk/Danzig, Königsberg, Pillau, and Memel.
130   Reichert, "Direct Supplies of Timbers", 135–138.

TABLE 15   Total number of Spain-bound ships from the southern Baltic recorded by the
            Sound Toll (1740–1799)

Period	Timber for naval construction	Timber for barrels	Grain	Other goods	Total
1740–1749	5	51	—	14	70
1750–1759	53	67	30	72	222
1760–1769	33	73	36	62	204
1770–1779	179	41	60	103	383
1780–1789	202	53	178	149	582
1790–1799	69	53	303	60	485
Total	541	338	607	460	1,946

SOURCE: REICHERT, "DIRECT SUPPLIES OF TIMBERS", 135

TABLE 16   Number of Spain-bound southern Baltic ships loaded with timber recorded by
            the Sound Toll (1740–1799)

Period	Gdańsk/ Danzig	Szczecin/ Stettin	Memel	Königsberg	Pillau	Other harbours	Total
1740–1749	5	—	—	—	—	—	5
1750–1759	41	10	—	1	—	1	53
1760–1769	16	7	3	7	—	—	33
1770–1779	66	20	59	19	7	8	179
1780–1789	12	10	144	21	9	6	202
1790–1799	12	2	47	5	3		69
Total	152	49	253	53	19	15	541

SOURCE: REICHERT, "DIRECT SUPPLIES OF TIMBERS", 137–138

their construction (see Table 16 and Map 5).[131] It is important to mention that
during this period—when there was an increase in the Spanish demand for
ship timber—ten ships-of-the-line and 14 frigates were built in the royal shipyards on the Iberian Peninsula (see Table 17).

---

131   For instance, in the 1776 timber inventory in the arsenal of La Carraca of 1776, masting and
       planking accounted for 35 per cent of the total. AGS, SMA, Arsenales leg. 355, Relación que
       manifiesta las Maderas para Construcción existentes en este Arsenal.

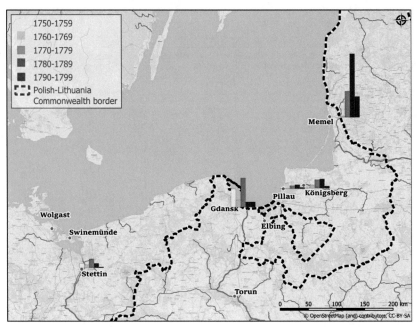

MAP 5 Timber shipped from the southern Baltic ports to Spain (1750–1799)

TABLE 17 Ships-of-the-line and frigates launched in the 1770s, during the northern timber *asientos* awarded to Chone and Marraci

Launching year	Ship's name	Shipyard	Type
1779	*Purísima Concepción*	El Ferrol	112-gun ship-of-the-line
1775	*San Eugenio*	El Ferrol	80-gun ship-of-the-line
1770	*San Pedro Apóstol*	El Ferrol	74-gun ship-of-the-line
1771	*San Pablo*	El Ferrol	74-gun ship-of-the-line
1771	*San Joaquín*	Cartagena	74-gun ship-of-the-line
1772	*San Juan Bautista*	Cartagena	70-gun ship-of-the-line
1772	*San Gabriel*	El Ferrol	74-gun ship-of-the-line
1773	Ángel de la Guarda	Cartagena	74-gun ship-of-the-line
1776	*San Dámaso*	Cartagena	70-gun ship-of-the-line
1779	*San Justo*	Cartagena	74-gun ship-of-the-line
1770	*Nuestra Señora del Rosario*	El Ferrol	34-gun frigate
1770	*Nuestra Señora del Carmen*	El Ferrol	36-gun frigate

TABLE 17   Ships-of-the-line and frigates launched in the 1770s, during the northern timber *(cont.)*

Launching year	Ship's name	Shipyard	Type
1771	*Santa Dorotea*	Cartagena	34-gun frigate
1772	*Nuestra Señora de la Asunción*	El Ferrol	36-gun frigate
1772	*Santa Perpetua*	El Ferrol	34-gun frigate
1772	*Santa María de la Cabeza*	El Ferrol	34-gun frigate
1773	*Santa Clara*	Cartagena	26-gun frigate
1773	*Santa María Magdalena*	El Ferrol	36-gun frigate
1774	*Santa Margarita*	El Ferrol	34-gun frigate
1774	*Santa Marta*	El Ferrol	34-gun frigate
1777	*Santa Rufina*	Cartagena	38-gun frigate
1777	*Santa Mónica*	Cartagena	34-gun frigate
1778	*Graña alias Nuestra Señora de la Paz*	El Ferrol	30-gun frigate
1779	*Santa Escolástica*	El Ferrol	30-gun frigate

SOURCES: RODRÍGUEZ GONZÁLEZ & COELLO, *LA FRAGATA EN LA ARMADA ESPAÑOLA*; GONZÁLEZ-ALLER HIERRO, *MODELOS DE ARSENAL DEL MUSEO NAVAL EVOLUCIÓN DE LA CONSTRUCCIÓN NAVAL ESPAÑOLA*, 237–303; BLANCO NÚÑEZ, *LA ARMADA ESPAÑOLA EN LA SEGUNDA MITAD DEL SIGLO XVIII*, 132; SÁNCHEZ BAENA & RODA ALCANTUD, "EL ARSENAL DEL MEDITERRÁNEO. CARTAGENA (1750–1824)", 117–198

6   Baltic Timber under the Control of the Banco de San Carlos and the *Asiento* of the Swedish Company Gahn

The formula by which contracts were awarded to Spanish merchants or Spain-based foreigners reached its maximum degree of development once the management of timber contracts was put under the Banco Nacional de San Carlos. The idea of creating a national financial institution to support the Crown's strategic projects emerged in 1779 on the initiative of the Count of Floridablanca and crystallized three years later when Charles Carlos III issued an order to create the bank, under the direction of Francisco Cabarrús. The bank's main aims were:

– providing financial support to the circulation of royal bonds and turning them into cash;
– supplying food and clothing to the armed forces;

– meeting the Crown's payments abroad;
– fighting economic speculation;
– providing credit for commercial and industrial ventures.[132]

The second of these aims was the most important, as it also involved the provision of timber. At first, the bank resorted to the traditional Riga and Saint Petersburg channels for supplies, through Marraci, Colombí, and, especially, Pedro Normande, who between 1781 and 1785 made several purchases of wood and hemp under the direct orders of the Secretary of State, which was closely watched by the Count of Floridablanca.[133] However, in 1786, the system of small *asientos* previously managed by Marraci, Colombí, Rey and Brandenburg, and Herman & Ellermann & Schlieper, as well as the direct purchases overseen by Normande, came to an end when the Banco de San Carlos assumed total control over supplies.[134] In fact, on 23 December 1786, the bank signed a contract to supply northern timber to the three naval departments for eight years (1787–1794). Like in Arragori's, Chone's, and Marraci's contracts, the agreement said that the timber "has to come from Riga or Saint Petersburg, freshly-cut, and free from sap and irregular knots".[135]

From the outset, the idea of centralizing the *asientos* in the Banco de San Carlos faced the opposition of the Cinco Gremios Mayores de Madrid, an association that represented the Spanish commercial elite. After supporting the Crown with loans during the American War of Independence, this association now demanded the *asientos* as payback. Cabarrús opposed this idea, but the failure of the bank to handle the supplies centrally forced him, in late 1788— barely two years after the bank signed the eight-year contract—to open negotiations with the Cinco Gremios, to which the business of supplying both the army and the navy was eventually granted for a 10 per cent profit.[136] A relation by the Swedish consul in Cádiz, Don Juan Jacobo Gahn (who collaborated with the bank), dated 11 November 1789, clearly reflects the difficulties involved in

132   Juan Hernández Franco, "Relaciones entre Cabarrús y Floridablanca durante la etapa de aquél como director del Banco Nacional de San Carlos (1782–1790)", *Cuadernos de historia moderna y contemporánea* 6 (1985): 82.

133   Rafael Torres Sánchez, "El estado fiscal-naval de Carlos III. Los dineros de la Armada en el contexto de las finanzas de la monarquía", in Juan Marchena Fernández and Justo Cuño Bonito (eds.), *Vientos de guerra. Apogeo y crisis de la Real Armada, 1750–1823*, vol. 1 (Madrid: Doce Calles, 2018), 427.

134   Valdez-Bubnov, *Poder naval y modernización del Estado*, 361.

135   "ha de ser de Riga o San Petersburgo, de corte más reciente, sin mucho sámago, ni de nudos irregulares". AGS, SMA, Asientos leg. 637, Condiciones bajo las cuales la Junta de Dirección del Banco Nacional de San Carlos propone a S.M. proveer de arboladura y tablazón.

136   Torres Sánchez, *Military Entrepreneurs*, 109–110.

meeting the demands of the three arsenals. According to this document, a lack of understanding of the northern timber business had driven the San Carlos to lose at least 100,000 pesos *duros* over the previous years.[137]

The change of management in naval supplies, from the bank to the Cinco Gremios, brought few advantages to the Bourbon state, as both Mediterranean and Baltic supplies remained firmly in the hands of foreign merchants. Cartagena, for instance, continued dealing with Baltasar Castellini, representative of the Italian society Monticelli,[138] and in the Baltic with the Cádiz-based Gahn company. The first contract awarded to this company for the supply of northern timber was signed in 1786, and from then onwards it held a virtual monopoly over the business.[139] The second contract was signed in 1791 and, according to a letter by Antonio Valdés, dated June 1794, the house received praise from the departmental committees: "They perform well in the supply of northern wood to the arsenals and the shipyard of Mahón".[140]

The decision to sign a new *asiento* was taken by the minister for the Navy based on the reports of the northern masting and planking delivered by Gahn to the naval departments. Interestingly, the timber earmarked for Cartagena in 1793 suffered a delay. The report triggered a reaction from the representative of the Swedish company in Cartagena, Juan de Benavente, who attached a list of the ships with origin in Russia to have moored in Britain and Copenhagen to avoid being captured in the Channel by French ships[141] when war broke out between Spain and Napoleonic France in March 1793.[142] Of the 20 ships listed by Benavente, 20 Spain-bound Dutch ships setting sail from Riga and Saint Petersburg with masting and planking, only four made it to their destinations in the summer of 1794. The rest were either captured or sold their timber to the British or the Danish, causing losses of 144,700 pesos to the *Real Hacienda*.[143] Despite these unexpected difficulties, the king approved a new contract with

137   AGS, SMA, Asientos leg. 637, El cónsul de Suecia Gahn.

138   Valdez-Bubnov, *Poder naval y modernización del Estado*, 361.

139   The Gahn company began cooperating with the navy in 1781, when it was awarded the asiento for the supply of copper for sheathing in La Carraca. Carrasco González, "Cádiz y el Báltico", 330–331.

140   "de buen desempeño en la provisión de maderas del Norte para las atenciones de sus arsenales y astillero de Mahón". AGS, SMA, Asientos leg. 637, Concediendo a don Juan Jacobo Gahn el asiento de provisión de arboladura y tablonería del Norte.

141   ANC, JED, libros de acuerdos vol. 2734, tomo 18 (1794).

142   John Lynch, *La España del siglo XVIII* (Barcelona: Crítica, 1991), 281–283; Jean René Aymes, *La guerra de España contra la Revolución Francesa (1793–1795)* (Alicante: Instituto de Cultura Juan Gil-Albert, Diputación de Alicante, 1991); Digby Smith, *The Napoleonic Wars* (London: Greenhill, 1998), 49–104.

143   Carrasco González, "Cádiz y el Báltico", 332–333.

Gahn, from 1 January 1795 to 31 December 1802.[144] This contract was signed against a turbulent background, because, after the war with France came to an end, Spain got involved in a series of conflicts with the British, during the so-called naval wars, between 1795 and 1805.[145]

Gahn also faced certain administrative and fiscal issues, because in several instances the cargoes included boards and other parts that had not been requested. Table 18 presents several examples in which La Carraca refused to accept cargoes that did not feature in the orders for 1794. When this happened, foreign dealers tried to sell the timber to private merchants, which involved paying taxes, so they often did their utmost to convince the customs officials that the cargoes were for the navy.[146]

TABLE 18   Ships with timber rejected by the intendent of La Carraca in 1794

Ship's name	Captain's name	Number of large boards	Number of boards	Price in reales de vellón
Concordia Dutch vessel	Simon Kooter	805	1,717	7,618.6
Concordia Dutch vessel	Rettie Janz Jangobert	149	1,072	3,155.17
Pietter Listra Dutch ketch	Teerben Bakker	320	786	2,338.17
Cathalina Dutch vessel	Voz	138 *berlingas*		243.30
Leenwik Dutch vessel	Tierlloff Vizer	262	677	4,718.11

SOURCE: AGS, SHH, LEG. 55, ESTADO COMPRENSIVO DE EXCLUIDA POR LA MARINA A LA CASA DE GAHN

---

144   AGS, SMA, Asientos leg. 637, Concediendo a don Juan Jacobo Gahn el asiento de provisión de arboladura y tablonería del Norte.

145   Cesáreo Fernández Duro, *Armada Española, desde la unión de los reinos de Castilla y Aragón*, t. VIII, cap. III–XV (Madrid: Sucesores de Rivadeneyra, 1895–1903); Glete, *Navies and Nations*, 277–278; Allan J. Kuethe and Kenneth J. Andrien, *The Spanish Atlantic World in the Eighteenth Century: War and the Bourbon Reforms, 1713–1796* (Cambridge: Cambridge University Press, 2014); José Gregorio Cayuela Fernández, "Marina española, estrategia conjunta y relaciones internacionales, 1713–1810", in Agustín Guimerá y Olivier Chaline (ed.), *La Real Armada y El Mundo Hispánico en el Siglo XVIII* (Madrid: UNED, 2022), 38–39.

146   AGS, SHH leg. 55, Estado comprensivo de excluida por la Marina a la casa de Gahn.

This abuse had constituted a problem since the Gil de Meesters' *asiento* in the 1750s,[147] which confirms Marquis González de Castejón's impressions on 16 February 1781:

> *asentista* Marraci never acts in bad faith in his contract with the king; this would be less of a big deal if it were not because previous *asentistas* never delivered in the quantities and dimensions they were asked for, so I expect that in the service of the king, that he will not be late in his deals.[148]

To stop these irregularities, on 12 July 1783 the king issued an order setting out toll duties for foreign timber, making provision for Spanish timber "to be moved from harbour to harbour within my dominions free from all charges".[149] Foreign timber not earmarked for the navy was subject to the following taxes:

- 4 maravedís de vellón per inch of Castilian *vara* of square timber from 3 inches thick and 3 inches wide to 8 and 9, inclusive, whose length is not over 10 *varas*;
- 5 maravedís de vellón per inch of Castilian *vara* of said thickness and width, 8 and 9 inches, over 10 *varas* in length;
- 5 maravedís [...] square timber over said thickness and width, 8 and 9 inches, not over 10 *varas* in length;
- 6 maravedís [...] square timber over said thickness and width, 8 and 9 inches, over 10 *varas* in length;
- 3 maravedís [...] boards and large boards;
- 2 maravedís [...] round timber;

---

147 For instance, a Dutch fluyt called *Jacoba Gertrudis* arrived in Cartagena on behalf of the Gil de Meesters in January 1752, theoretically loaded with mainmasts and foremasts for 70-gun ships-of-the-line. After reviewing the cargo, the captain of the maestranza, Don Martín de Esparza, and the masting master, Don Eduardo Bryant, realized that the cargo was of little use because the measurements did not match the order. The intendent of the department in Cartagena, Don Francisco Barrero Peláez, in a letter requested that the minister for the Navy "order [the timber] to be more to specifications, and for the right type [as per the *asiento*] to be delivered in time for the [70-gun] ships to be built in this harbour". AGS, SMA, Arsenales, legajo 612.

148 "al asentista Marraci nunca se le faltará la buena fe de lo contratado con el rey; falta que no sería tan grande, si los anteriores asentistas, que jamás cumplieron en cantidades calidades ni dimensiones de lo que se les encargó, no lo hubieran originado y así espero que por lo que interesa el servicio del Rey y que no se atrase en cuanto sea posible en estas conducciones". AGS, SMA, Arsenales leg. 635, El intendente envía copia certificada de crédito despachado a favor de Marraci.

149 "de unos puertos a otros de estos Dominios con libertad de derechos".

- 16 maravedís [...] oars from The Hague, and 8 maravedís those of Flemish pine;
- 5 maravedís [...] *duelas* from Hamburg and other northern parts for large barrels;
- 4 maravedís for half barrels and 3 for barrels;
- 4 maravedís de vellón per inch of Castilian *vara* for *duela blanquilla* from Italy, Virginia, and other parts, of dressed or undressed oak or chestnut tree, for barrels; 3 maravedís for half barrels; and 2 maravedís for casks.[150]

However, as illustrated by the example of Gahn's ships, the practice of bringing timber that had not been ordered could not be eradicated. This is also shown by the lawsuits filed by Cádiz customs officials and some of the captains arriving with these cargoes in Spain. However, custom officials, using the powers granted by the Ordinance of 12 June 1773, charged taxes to the wood heading for La Carraca, as shown by a letter by Diego de Gardoqui, signed on 17 July 1794. In response to the complaints filed by freighters, he passed information from the Ministry of Finances to the general superintendent of Cádiz:

> the *Junta* of the Navy Department in Cádiz reached an agreement after analysing the damage suffered by foreign wood suppliers to the shipyard of La Carraca because the customs offices charge them the full duties if they do not demonstrate that their cargoes are for the construction and repair of the king's ships.[151]

This conflict between the Cádiz customs, foreign *asentistas*, and officials in La Carraca remained unresolved for years. Navy authorities tried to support the ship skippers, but admitted that they could not give a specific answer about the use to which the timber would be put in advance, emphasizing that "the Department cannot certify the use until [the timber] has been examined, which could take many years, as some timber needs to season for a long time, and not used immediately".[152] Juan Jacobo Gahn, who had been working with the *asientos* since 1786, was given, 11 years later, a guarantee from the Cádiz custom office

---

150   Archivo Histórico Provincial de Cádiz (AHPC), Real Aduana, vol. 38, fs. 294–303v.

151   "la Junta del Departamento de Marina de Cádiz formó un acuerdo en que después de tratar de los perjuicios que sufren los Proveedores de Maderas extranjeras para servicio del Arsenal de La Carraca con motivo de obligárseles en aquella Aduana a depositar el total importe de los derechos de las mismas maderas mientras no acreditasen su consumo en la construcción o habilitación de Bajeles". AHPC, Real Aduana, vol. 46, fs. 14–15.

152   "las oficinas del Departamento no pueden dar certificación de la inversión hasta que se verificase [la madera], lo cual podría tardar en suceder muchos años, ya por los que

that no *alcabala* or any other charge was to apply to the masting and planking brought in the context of the eight-year contract in operation from 1 January 1795.[153] Rey and Brandenburg—which, as noted, also participated in the supply of timber from the 1790s—faced the same problem between March and August 1792, with a cargo comprising 135 oak beams, 50 large boards, 36 *baos*,[154] and 36 *curvas*[155] brought to Cádiz by the Dutch vessel *Dregterland*. Josef Rodríguez, the house's representative, handed over his manifesto to the intendent of the Navy in La Carraca, and the customs officers did not charge him any taxes. However, when the cargo was examined, it transpired that some of the parts in the hold were not in the arsenal's order, so customs payment was requested for those. The lawsuit dragged on until 10 August 1792, when Rodríguez agreed with the department accounts office to buy the unwanted timber.[156]

This and other examples, going back to the times of the Gil de Meesters, show that the timber business was a lucrative trade. Foreign merchants and ship captains tried, in addition, to carry out semilegal deals under the pretext that the timber was earmarked for the *asiento*, when it was not always the case. The Crown tried to put an end to these irregularities issuing orders such as that in June 1783, setting up customs payments for timber that was not for the Navy.[157] The order also reflects Charles III mercantilist policies to defend the national interests, as Spanish ships were made exempt from these payments.[158]

Finally, it is important to emphasize that, when the Banco San Carlos took over the supply contracts with the Ministry of the Navy, a change of secretary also took place, after the death on 19 March 1783 of Marquis González de Castejón, who was replaced by Antonio Valdés y Bazán. Valdés was concerned about the enormous expenses posed by maintaining and modernizing the navy and tried to change the funding model. His proposal suggested paying for ship construction with the revenue generated by the tobacco monopoly in Cuba. The proposal was rejected because of the opposition of politicians and merchants involved in this business, and the funding of shipbuilding remained the responsibility of the *Real Hacienda*. In fact, the main obstacle is that the

---

necesitan algunas de las maderas para preparase y curarse, por lo que se suele tardar en emplearlas". AHPC, Real Aduana, vol. 46, fs. 14.

153   AHPC, Real Aduana, vol. 46, f. 265.
154   Eng. beam.
155   Eng. knee.
156   AHPC, Real Aduana, vol. 42, fs. 208–213v.
157   Ibid., vol. 38, fs. 294–303v.
158   Lynch, *La España del siglo XVIII*, 193–194.

TABLE 19   Annual maintenance expenses of the *Marina Real* after the American War of
            Independence

Year	Expenses
1783	135,000,000 reales de vellón
1784	136,000,000 reales de vellón
1785	127,000,000 reales de vellón
1786	131,000,000 reales de vellón
1787	147,000,000 reales de vellón

SOURCE: MUSEO NAVAL DE MADRID (MNM), MS 471, FS. 32–33V

tobacco monopoly did not raise enough revenue to meet the Navy's financial
needs.[159]

The cost of keeping the navy on a war footing both in Spain and America led
to a massive public debt crisis, and from 1793 onwards the Navy's funding needs
were seen as a severe problem, and underfunding became a chronic problem,
as reflected in the scarcity of raw materials such as wood, hemp, and canvas.
This resulted in the decline of Spanish naval power during the so-called naval
wars of the late 18th and early 19th centuries.[160] Antonio Valdés y Bazán (1783–
1795) was the last minister to preside over a relatively efficient and stable navy;
he closed a period of statesmen who, beginning with Ensenada, managed to
modernize the Spanish navy and turn the country into the second naval power
in the world.

---

159   Jacques A. Barbier, "Indies Revenues and Naval Spending: The Cost of Colonialism for
       the Spanish Bourbons, 1763–1805", in Jeremy Black and Frederick C. Schneid, *Warfare in
       Europe 1792–1815* (London: Routledge, 2007), 14–15.
160   Barbier, "Indies Revenues and Naval Spending", 12; Valdez-Bubnov, *Poder naval y modern-
       ización del Estado*, 414–415.

The Southern Baltic Hinterland and Timber Extraction

Chapter 1 defined the main outlines of the timber supply policies adopted by the Spanish Crown, represented by the Ministry of the Navy and the Indies, in the second half of the 18th century, and analysed the most important contracts signed by the Spanish state with domestic and foreign entrepreneurs for the supply of timber from the main southern Baltic harbours, such as Riga, Memel, Königsberg, Danzig/Danzig, and Szczecin/Stettin.

These contracts offer an overview of the volume and type of trade between the southern Baltic and the Spanish maritime departments. Chapter 2 presents information about timber extraction in the hinterlands of the ports of Riga, Memel, Königsberg, Danzig/Danzig, and Szczecin/Stettin. These harbours were connected to the forests of the Kingdom of Prussia and the Polish-Lithuanian Commonwealth by navigable rivers, down which the felled timber was floated. The chapter also provides interesting information about the forestry policies introduced by the king of Prussia, the king of Poland, and the magnates who owned forest areas in their estates.

1   Timber Cutting in Pomerania and along the Odra River and Its Trade in Szczecin

In 1713, Szczecin/Stettin was taken over by the Prussians and, after the Treaties of Stockholm, on 21 January 1720, the city, together with the Swedish part of Pomerania, was officially sold for 2 million thalers to the king of Prussia, Frederick William I. In the following years, the Prussian authority consolidated in the port and its hinterland.[1] One of the most important achievements of the first half of the 18th century was the construction of new fortifications that replaced the medieval defensive walls. Traditionally, Szczecin/Stettin, as the dominant port in the region, focused its commercial activities on the exchange of goods with Wolgast, Greifswald, and Stralsund, and, less frequently, with

---

1 Petri Karonen, "Coping with Peace after a Debacle: The Crisis of the Transition to Peace in Sweden after the Great Northern War (1700–1721)", *Scandinavian Journal of History* 33 (2008), 203–225.

© KONINKLIJKE BRILL BV, LEIDEN, 2024 | DOI:10.1163/9789004689640_004

Rostock, Wismar, and Lubeck. Other important places for Szczecin's trade were Copenhagen, Hamburg, Amsterdam, Rotterdam, The Hague, Ostend, Antwerp, Bordeaux, and London. Among other commodities, ships from these ports came to Szczecin/Stettin loaded with colonial products (coffee, tea, cocoa, indigo, spices), herring, liqueurs, wine, brandy, olive oil, cotton, and manufactured goods. In exchange, Szczecin/Stettin sent grain, linseed, salt, stone for construction, cloth, and large quantities of wood for the shipbuilding and construction industries as well as staves, barrel bottoms, and firewood.[2]

The natural base from which the wood on sale in the city came were the surrounding forests, located within a 50-km radius around the Szczecin Lagoon, whence timber was hauled by road and sold on the timber market (Ger. *Holzmärkte*). According to the *Wöchentlich-Stettinische Frag und Anzeigungs-Nachrichten*, 86 logging sites were active in the jurisdiction of Pomerania in the period 1750–1771. The main logging stations were the villages of Podejuschen (with 30 mentions concerning timber-related transactions), Szczecin/Stettin (20 mentions), Gollnow (17 mentions), Saatzig (12 mentions), Hinzendorf (11 mentions), Pudagla (11 mentions), Damm (11 mentions), Friedrichswalde (ten mentions), Ueckermünde (ten mentions), Nipperwiese (seven mentions), Grambin (six mentions), Dunzig (six mentions), and Moritzfelde (five mentions). Other locations within this jurisdiction feature in the record between one and three times (see Map 6).[3]

---

2 Henryk Lesiński, "Handel morski Szczecina w okresie szwedzkim 1639–1713", *Materiały Zachodniopomorskie* 31 (1985), 277–295; Radosław Gaziński, *Handel morski Szczecina w latach 1720–1805* (Szczecin: Wydawnictwo Naukowe Uniwersytetu Szczecińskiego, 2000), 273–277; Józef Stanielewicz, "Zarys rozwoju portu i handlu morskiego Szczecina od XVI do XVIII wieku", in Paweł Bartnik and Kazimierz Kozłowski (eds.), *Pomorze Zachodnie w tysiącleciu* (Szczecin: Wydawnictwo ap, 2000), 121–128; Henryk Lesiński, "Przemiany w stosunkach handlowych miast Pomorza Zachodniego w drugiej połowie XVII i początkach XVIII wieku", in Gerard Labuda (ed.), *Historia Pomorza (do roku 1815)* vol. 2 (Poznań: Wydawnictwo Poznańskie, 1984), 173–216; Bogdan Wachowiak, "Wybrane problemy handlu warciańsko-odrzańskiego w latach 1618–1750", *Przegląd Zachodniopomorski* 26 (2011), 49–66; Michał Knitter, "Verifizierung von Schifffahrtsstatistiken des Stettiner Hafens in der zweiten Hälfte des 18. und Aufgang das 19. Jahrhunderts", *Studia Maritima* 25 (2012), 23–51; Emil Chróściak "Szczecin's Maritime Timber Trade and Deliveries to Spain between 1750 and 1760 on the Basis of Wochentlich-Stettinische Frag- und Anzeigungs-Nachrichten", *Studia Maritima* 33 (2020), 150–164.

3 Emil Chróściak "Szczecin's Maritime Timber Trade and Deliveries to Spain", *Studia Maritima* 33 (2020), 152–153. Książnica Pomorska, Szczecin (KPS), *Wöchentlich-Stettinische Frag*, the reference number XVIII.15123.I/10,ST6947, for the period 1761–1771. These places were: Jasenitz, Kolbatz, Neugard, Bahn, Rothenfier, Greifenberg, Kaseburg, Pyritz, Mühlembeck, Anclam, Gotzlow, Lauenburg, Marienfließ, Marsdorf, Messenthin, Pölitz, Schwedt, Stolp, Stecklin, Wildenbruch, Schivelbein, Bärwalde, Basenthin, Bütow, Freienwalde, Temnick, Neuenhagen,

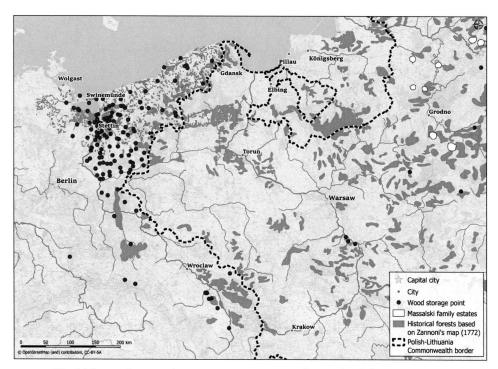

MAP 6   Wood felling and storage places in Pomerania mentioned in *Wöchentlich-Stettinische*
SOURCE: LOOK4GIS-LUKASZ BRYLAK BASED ON QGIS SOFTWARE

Table 20 presents only adverts targeted at the shipbuilding industry. It must be recalled that this type of wood only accounted for approximately 13 per cent of all the timber channelled through Szczecin/Stettin in the 1750s. The largest market segments were firewood (over 25 per cent), barrel staves and bottoms (about 25 per cent), unidentified purposes (about 23 per cent), and timber for the construction industry (13 per cent). It is important to note that during the period under consideration, Copenhagen was the main destination for ships departing from Szczecin/Stettin and Swinemünde (a satellite harbour for Szczecin/Stettin); 69.1 per cent of all ships registered as carrying wood arrived

---

Ferdinandstein, Rügenwalde, Jungferberg, Ganserin, Gülzow, Greifenhagen, Ihnamünde, Nörenberg, Gerzlow, Körlin an der Persante, Klütz, Kublitz, Köslin, Köstin, Krangen, Krampe, Cremzow, Hackenwalde, Kranzin, Kucklow, Külz, Leopoldshaben, Holzhagen, Matzdorf, Massow, Bärwalde in der Neumark, Münchendorf, Barenbruch, Neuwarp, Pasewalk, Plathe an der Rega, Pranzlau, Pribbernow, Kuhdamm, Retzowsfelde, Redel, Reckow, Nassenheide, Groß Karzenbur, Pollychen, Rehdorf, Karlsbach, Neustettin, Falkenwalde, Thurow, Stargard, Stolpmünde, Warsow, Fiddichow, Wollin, Wittstock, Finkenwalde, Schwerin, and Züllchow (see Map 6).

TABLE 20   Adverts for the sale of ship timber in Szczecin/Stettin according to the *Wöchentlich-Stettinische Frag* (1750–1771)

Year	Wood for shipbuilding	Origin
1750	1,198 pieces of shipbuilding oak wood	Gützow
1751	25 oak logs for shipbuilding	Ueckermünde
	500 fathoms of shipbuilding wood	Regalitz
	94 planks of oak wood	Podejuschen
1752	shipbuilding wood	Wittstock
	shipbuilding wood	Ueckermünde
	shipbuilding wood	Neuhaus
	shipbuilding wood	Friedrichswalde
	60 pieces of shipbuilding oak wood	Pudagla
	30 planks of oak wood	Friedrichswalde
	50 planks of oak wood	Massow
1753	oak shipbuilding wood	Bahn
	oak shipbuilding wood	Moritzfelde
1754	—	—
1755	200 fathoms of shipbuilding oak wood	Damm
	25 pieces of shipbuilding oak wood	Podejuschen
1756	shipbuilding wood	Jasenitz
	shipbuilding wood	Ihnamünde
	48 pieces of shipbuilding wood	Podejuschen
	174 pieces of ship curves	Lantz
1757	shipbuilding wood	Leuenburg
1758	50 pieces of shipbuilding oak wood	Mühlenbeck
1759	shipbuilding oak and spruce wood	Damm
1760	—	—
1761	300 planks of oak wood and shipbuilding wood	Stettin
	49 oaks, 9 large spruce beams, 14 large spruce sail blocks	Rothenvier
1762	650 oaks for shipbuilding	Lippehne
1763	480 fathoms of shipbuilding beech wood, 1,135 fathoms of shipbuilding oak wood, and 5,400 fathoms of shipbuilding spruce wood.	Royal Forests of Stettin, Ueckermünde, Pudagla and Wollin
	302 oaks for shipbuilding, 258 fathoms of shipbuilding beech wood, 2 fathoms of shipbuilding spruce wood, 510 fathoms of shipbuilding alder wood	Royal Forests of Friedrichswalde, Colbatz, Stepenitz, Göltzow, Neugardten, Saatzig, Rügenwalde and Bütow
		Anclam

TABLE 20   Adverts for the sale of ship timber in Szczecin/Stettin according (*cont.*)

Year	Wood for shipbuilding	Origin
1763	188 pieces of shipbuilding oak wood	Ganserin
	12 ship curves and 100 planks	Gollnow
	200 oaks and 700 fathoms of shipbuilding alder wood	Podejuschen
	9 pieces of shipbuilding oak wood	
1764	2 and 2.5 inch oak planks and small shipbuilding wood	Podejuschen
	1,060 fathoms of shipbuilding beech wood, 1,190 fathoms of shipbuilding oak wood, 2,160 fathoms of shipbuilding alder wood 8,870 fathoms of shipbuilding spruce wood	Royal Forests of Stettin, Ueckermünde, Pudagla and Wollin
	400 oaks and 300 beech	Forest of Captain Woedtcke
1765	—	—
1766	256 pieces of shipbuilding oak wood	Graf's Podewils forests near the village of Wusterwitz
	50 pieces of shipbuilding oak wood	Rügenwalde
	50 pieces of shipbuilding oak wood and 30 pine trees for masts	Bütow
	fourteen pine trees for masts and 4,064 shipbuilding timbers of various kinds	Stettin
	200 fathoms of shipbuilding alder wood	Gollnow
1767	nine pieces of shipbuilding oak wood	Podejuschen
	36 spruce deck planks, 36 feet long	Stettin
1768	100 pieces of shipbuilding oak wood	Rügenwalde
	50 pieces of shipbuilding oak wood and 30 pine trees for masts	Bütow
	shipbuilding wood	Woldenberg
1769	shipbuilding oak wood and planks	Stettin
	581 pieces of shipbuilding wood	Stargard
1770	200 oak logs for planks and boards	Glogau
1771	shipbuilding wood	Friedrichswalde
	six pieces of ship curves	Rothenvier

SOURCE: KPS, *WÖCHENTLICH-STETTINISCHE FRAG*, XVIII.15123.I/10,ST6947, FOR THE PERIOD 1750–1771

in Copenhagen. Other important destination harbours for Pomeranian timber were London (7.1 per cent), Bordeaux (5 per cent), and Amsterdam (4.5 per cent). The remaining 14.3 per cent of ships carrying timber set sail for other European and Baltic ports.[4]

Back to the issue of shipbuilding timber on offer in Szczecin/Stettin, it is worth noting that it came from various sources: it was sold at the timber market by merchants, *Junkers* (landowners), monks from the monastery of Saint John in Podejuschen, and the officials who managed the royal forests near Szczecin/ Stettin, Ueckermünde, Pudagla, Wollin, Friedrichswalde, Colbatz, Stepenitz, Göltzow, Neugardten, Saatzig, Rügenwalde, and Bütow.[5] Most of the wood they offered to the shipbuilding industry is generally defined as *schiffholz* (shipbuild- ing wood). Occasionally, the exact use of the timbers for masts,[6] curves,[7] and hull planks is also mentioned.[8] In some instances, the specific type of wood is also mentioned, mostly oak, but also beech, alder, spruce, and pine. In addi- tion, not all adverts specify the exact amount of wood sold, while others refer to pieces, logs, beams, and trees.[9] All of these confirm that Szczecin/Stettin, with its forest hinterland in Pomerania, was a major supplier of wood for the ship- building industry. Unfortunately, internal market data from Szczecin/Stettin do not specify exactly to whom the wood was delivered, but it can be assumed that a major part was intended for export, which is confirmed by the maritime traf- fic registered in the *Wöchentlich-Stettinische Frag* during the period 1750–1771.[10]

Apart from the Pomeranian forests, the newspaper also advertised wood from the Province of Neumark or the New March, whence it was shipped to Szczecin/Stettin in two ways: pulled by oxen by overland routes and in the form of logs rafts pushed down the Oder River and the Warta River (a major tribu- tary of the Oder). The place that features in the record the most times between 1759 and 1770 was Cartzig (eight mentions). Moreover, wood from other loca- tions—e.g. Cladow, Massin, Mückenburg, Staffelde, Neuhaus, Regenthin, and Schlanow—was also often on offer (Table 21).[11]

---

4   Emil Chróściak "Szczecin's Maritime Timber Trade and Deliveries to Spain", *Studia Maritima* 33 (2020), 155–157.

5   KPS, *Wöchentlich-Stettinische Frag*, XVIII.15123.I/10,ST6947, for the period 1750–1771.

6   For example, in 1766, 14 pine trees for masts were on sale in the Szczecin/Stettin market (ibid.).

7   For example, in 1763, 12 ship curves and 100 planks from Ganserin were on sale (ibid.).

8   For example, in 1751, 94 planks of oak wood from the monastery of Podejuschen were on sale (ibid.).

9   Ibid.

10   Ibid.

11   Ibid., for the period 1759–1770. Towns and villages of Neumark, which are listed in *Wöchen- tlich-Stettinische Frag*, from which timber was shipped to Szczecin/Stettin: Bischofsee, Braschen, Wartenberg, Schwachenwald, Zicher, Zachow, Driesen, Drewitz, Görlsdorf,

TABLE 21   Adverts for the sale of shipbuilding wood from the Province of Neumark in Szczecin/Stettin according to the *Wöchentlich-Stettinische Frag* (1759–1770)

Year	Wood for shipbuilding	Origin
1759	40 oak logs for planks	Stabenow
	40 oak logs for planks	Schwachenwald
	200 oak logs for planks	Regenthin
	40 oak logs for shipbuilding wood	Hammer
	20 oaks logs for shipbuilding wood, 20 oaks for planks, 100 Stettin pines,	Schlanow
	12 ship masts, 300 Stettin pines	Mückenburg
1761	20 pine ship masts	Cartzig
	20 pine ship masts	Cladow
	20 pine ship masts	Regenthin
1762	six pine ship masts	Regenthin
	20 oaks logs for shipbuilding wood	Sellnow
	25 oaks logs for shipbuilding wood	Schwachenwald
	16 pine ship masts	Cladow
	20 oaks logs for shipbuilding wood	Himmelstädt
	20 oaks logs for shipbuilding wood	Neumühl
	ten pine ship masts	Cartzig
	20 oaks logs for shipbuilding wood and six pine ship masts	Staffelde
	20 oaks logs for shipbuilding wood	Braschen
	30 oaks logs for shipbuilding wood and ten pine ship masts	Driesen
	20 oaks logs for shipbuilding wood and 24 pine ship masts	Schlanow
	20 oaks logs for shipbuilding wood	Hammer
1763	ten ship masts	Cartzig
	ten ship masts	Mückenburg
	ten ship masts	Neuhaus
	ten ship masts	Driesen
	25 ship masts	Schlanow

Cartzig, Cladow, Neuhaus, Leine, Himmelstädt, Massin, Neumühl, Neuenburg, Regentlin, Reppen, Mückenburg, Schlanow, Staffelde, Stölpchen, Züllichow, Schönfließ, Wildenow, Stabenow, Sellnow, Braschen, Lippehne, Berlinchen, Landsberg an der Warthe, Friedeberg, Zielenzig, Custrin, and Vietz.

TABLE 21   Adverts for the sale of shipbuilding wood from the Province of Neumark (*cont.*)

Year	Wood for shipbuilding	Origin
	ten ship masts	Massin
	50 ship masts	Cladow
	ten ship masts	Regenthin
1764	six ship masts	Cartzig
	ten ship masts	Mückenburg
	six ship masts	Neuhaus
	six ship masts	Cladow
	six ship masts	Regenthin
1765	ten pine ship masts	Cartzig
	ten pine ship masts	Neuhaus
	six pine ship masts	Staffelde
	ten pine ship masts	Mückenburg
	six pine ship masts	Regenthin
	six pine ship masts	Massin
	six pine ship masts	Cladow
1766	ten pine ship masts	Schlanow
1767	ten ship masts	Neuhaus
	eight ship masts	Staffelde
	eight ship masts	Driesen
	20 ship masts	Schlanow
	ten ship masts	Massin
1768	ten ship masts	Cartzig
	ten ship masts	Neuhaus
	ten ship masts	Mückenburg
	ten ship masts	Driesen
	ten ship masts	Wildenow
1769	six ship masts	Cartzig
	six ship masts	Neuhaus
	eight ship masts	Staffelde
	four ship masts	Driesen
	12 ship masts	Massin
	six ship masts	Cladow
	eight ship masts	Himmelstädt
	eight ship masts	Wildenow
	six ship masts	Reppen
	eight ship masts	Linichen

TABLE 21   Adverts for the sale of shipbuilding wood from the Province of Neumark (*cont.*)

Year	Wood for shipbuilding	Origin
1770	six ship masts	Neuhaus
	six ship masts	Cartzig
	six ship masts	Staffelde
	five ship masts	Mückenburg
	four ship masts	Driesen
	six ship masts	Schlanow

SOURCE: KPS, *WÖCHENTLICH-STETTINISCHE FRAG*, XVIII.15123.I/10,ST6947, FOR THE
PERIOD 1759–1770

The adverts for shipbuilding timber from the forests in the Province of
Neumark show that wood from there was mainly used for masts, a trend confirmed by adverts from the period 1763–1770, in which this type of product pre-
dominates, in contrast with the earlier period (1759–1762), during which the
products on offer include an assortment of masts and oak logs and planks.[12]
However, it is important to note that the wood on offer also included other
timber products that may have also been used by the shipbuilding industry,
but whose exact purpose was not made clear in the adverts. The adverts often
mention masts, for example, "200 pines, and twenty beech trees"[13] or "40 oaks,
ten masts, 400 pines",[14] which may suggest that the entire cargo was intended
for the shipbuilding industry. Cargoes coming through Szczecin/Stettin from
the forests of Pomerania and the New March strongly suggest that the latter
jurisdiction was mainly involved in mast production. This is indicative of the
specialised nature of forest productions there, which focused on good quality
pine trees, the species most often used to build spars.

Finally, timber rafts were also sailed down the Oder River[15] from the
Duchies of Silesia—which were annexed by Prussia after the First Silesian
War (1740–1742)[16]—with major storage sites in Ratibor, Flinsberg, Petersdorf,
Grase, Glogau, Mangersdorf, Tyllowitz, Jägerhäuß, Schedyske, Schiedlow, and

---

12   Ibid., for the period 1750–1771.
13   Offer to sell wood from the Staffelde forests (Ibid., 1763, Nr. 26, 437–438).
14   Offer to sell wood from the Massin forests (Ibid., 1768, Nr. 26, 550–551).
15   Wachowiak, "Wybrane problemy handlu warciańsko-odrzańskiego", 49–66.
16   Robert B. Asprey, *Frederick the Great: The Magnificent Enigma* (New York: Ticknor and
      Fields, 1986), 177–181; David Fraser, *Frederick the Great: King of Prussia* (London: Allen
      Lane, 2000), 120–121.

Stoberau.[17] Stoberau was the headquarters of timber rafting down the Stoberau River into the Oder. In 1766, royal official Frick reported a shipment of approximately 3,000 cubic metres of ship timber from the forests of Dammerau towards Szczecin/Stettin. The purpose of this wood is unknown, but this and earlier reports from 1748, 1754, and 1755 confirm that this Pomeranian harbour received wood for different purposes, including the shipbuilding industry.[18] The source of wood in Lower Silesia was the Kłodzko Valley (Ger. *Glatzer Kessel*), where the royal forests of Reusalzer, Wohlauer, Herrnstadt, Wohlau, and Neusaltz supplied durable alder and oak wood and also pine and spruce.[19]

2   Examples of Forest Management by the Prussian Crown in Silesia, Kłodzko, and Pomerania in the Second Half of the 18th Century

It is worth emphasising that the Prussian Crown also carefully exploited its forest resources. This is illustrated by the royal order entitled *Holtz, Mast und Jagd Ordnung für Unser Souverainen Hertzoghum Schlesien und die Souveraine Grafschafft Glatz*, issued by Frederick II the Great on 2 December 1750,[20] in which the monarch presented a series of important forest management rules. The royal decree consists of 26 laws, which are also broken down into sections. The laws deal with the management of forests under the jurisdiction of cities and villages, but also that of the Military and Treasury Cameras in Wrocław and Głogów, which were directly concerned with the protection and cultivation of forests. The latter aspect was addressed by the first paragraph, entitled "Von denen Holzungen und Schonung auch Anbauung derselben", which contains 13 points and is the most comprehensive in the entire royal order. The first two points put the felling of trees under state control and forbid timber harvesting beyond what was strictly necessary from an economic perspective. The next point forbids the sale of shipyard oak wood, boards for barrels, and other forms of cooperage unless a permit issued by the Silesian Military and Treasury Cameras was secured. The sale of wood from episcopal,

---

17   Archiwum Państwowe Opole (APO): reference numbers: 45/1191/0/5/497, 45/1191/0/5 /4987, 45/1191/0/5/4972, 45/1242/0/16.1/4786, and kps *Wöchentlich-Stettinische Frag*, XVIII. 15123.I/10,ST6947, for the period 1750–1771.

18   APO, 45/1191/0/5/4972, "Holz-Flössung Gerechtigkeit des Dom. Domerau auf dem Stoberbache".

19   APO, 45/1242/0/16.1/4786, "Regulatif nach welchem die Königlichen Domainen- Forsten in Schlesien behandelt".

20   Archiwum Państwowe Wrocław (APW), Hochberg, exp. 1419, fs. 1–35.

monastic, or parish jurisdiction with a value of over 200 Rhenish thalers also had to be reported and supervised by the Military and Treasury Cameras. The Crown ordered owners of private forests to keep one-tenth of their forests in reserve and to afforest clearings. In point 10, the Crown prohibited the free grazing of livestock, especially goats, which ate young tree shoots and slowed down the recovery of forests. The Crown allowed for controlled deforestation for cultivation, but after the soil was depleted, these areas were to be reforested. Finally, in relation to the last point, the king promoted the afforestation of mountain slopes in the Duchy of Silesia and the Kłodzko County, especially in those places and areas where wood was a valued product and forests grew slowly.[21] The issue of this royal edict in 1750—i.e. eight years after the incorporation of Silesia and the territory of Kłodzko into Prussia—shows the scale of the destruction of local forests and the previous lack of control over the wood harvested. The introduction of forestry supervision agencies, especially the Military and Treasury Cameras in Wrocław and Głogów, reduced uncontrolled logging for firewood and stimulated afforestation, especially in mountain areas of Upper Silesia. In the years 1763, 1773, 1777, 1778, 1782, and 1785, further royal decrees were published, but in many points, they merely reproduced the first edict of Frederick II in 1750.[22] It is worth noting that the king was also interested in communal areas, and from 1754 state foresters were entrusted with the supervision of communal forests, in which they were to earmark trees to be cut down and harvested; however, they did not receive a salary for this additional task, which led to these areas being poorly supervised.[23]

In Pomerania, the Prussian kings also took care of securing forest resources and pursued a policy of afforestation. Between 1777 and 1783, several edicts were published concerning the issue of forest management. The first two were published on 24 December 1777. The first forest ordinance for Pomerania established the principles for felling, selling, taxing, and delivering wood, as well as the lease of moors (Ger. *Forstordnung für Pommern von Anweisung, Fällen, Verkauf, Taxe, und Lieferung, des Holzes, auch Einmiethen in der Heide*).[24] The second document contained detailed instructions for forest and hill

---

21   APW, Hochberg, exp. 1419, fs. 3–8.
22   Aleksander Nyrek, *Gospodarka leśna na Górnym Śląsku od połowy XVII do połowy XIX wieku* (Wrocław: Zakład Narodowy im. Ossolińskich, 1975); Aleksander Nyrek, "Stan i praktyki wiedzy leśnej na Śląsku do połowy XIX wieku", *Śląski Kwartalnik Historyczny Sobótka* 3 (1972), 427.
23   Bernhard E. Fernow, *A Brief History of Forestry in Europe, the United States and Other Countries* (Toronto: Toronto University Press, 1911), 56.
24   Archiwum Państwowe Szczecin (APS), Akta pomorskich superintendentur, reference numbers: 65/34/0/-/5, fs. 26–31.

management (Ger. *Forstordnung für Pommern von Bearbeitung der Forstsachen und Verwaltung der Forstgefälle*).[25]

Both edicts from 1777 contain instructions about forest cleaning and care-taking (e.g. ordering the removal of broken branches), and orders for forest officials to stay in logging areas. Interestingly, the instruction says that wood could be collected from whole forested areas, because the division of forests into different zones, which would later help control logging activities, had not yet been fully introduced. Although wood could be sold all year round, during fallow periods of, for example, oak and beech, only the cutting down of dry and sick trees was allowed. The harvesting of wood worth more than 20 Rhenish thalers required a licence. Below this amount, it could be harvested without a licence, but the forestry office had to be notified about the felling. Interestingly, cutting down young pines, spruces, and firs for boards was forbidden. How-ever, trees blown over by the wind, dead, or sick could be used for this purpose. If there was not enough wood for planks, they could be sourced from other species such as juniper, birch, aspen, and alder.[26] The king ordered that these species be used as a substitute for pines, spruces, and firs in construction. The instruction *Forstordnung für Pommern von Anweisung, Fällen, Verkauf, Taxe, und Lieferung, des Holzes, auch Einmiethen in der Heide* ended with a detailed table of the taxes to be collected by the Cameras in Pomerania. Table 22 pres-ents examples of wood batches for shipbuilding, largely intended for export from the port of Szczecin/Stettin.[27]

From the remaining royal edicts, it is worth paying some attention to the *Instruction* for laying clumps of acorns and, in general, on the improvement of forest management for better cultivation of oaks in the district of Łeba, pub-lished on 25 January 1780.[28] The edict emphasises the importance of oak for domestic, construction, and industrial uses. The text points out that despite the widespread use and high demand for oak wood, no regulations had been passed to that date to better manage this type of forest. The manual focuses on five main topics: the right time to collect acorns; storage conditions for acorns; selection of appropriate land for sowing oaks; guidance on sowing techniques; and transplanting. The document suggests, for example, that acorns are col-lected when a large number of them begin to fall, but that they are not collected

---

25   Ibid., fs. 2–7.
26   Ibid., fs 2–7 and 26–31.
27   Ibid., fs. 30–31.
28   APS, Akta miasta Łeby, exp. 65/205/0/6/28, fs. 41–47. Anweisung zum Anlegen der Eichel-Kämpe, und überhaupt zum besseren Fortbringen des so nützlichen als unentbehrlichen Eich-Baums.

TABLE 22   Examples of tax fees for ship timber and masts from the 1777 edict

Tax on timber and masts for sale in the country	Western Pomerania (*Vorpommern*)		Pomeranian (*Hinterpommern*)	
	Rhenish thalers	Pennies	Rhenish thalers	Pennies
A shock of 64 pieces of oak ship nails	—	8	—	8
A shock *Bootsinholz* of 64 pieces, each piece up to 12 feet long and 3–5 inches square	12	—	10	—
A large ship's mast from 70 to 84 feet, 18–20 inches in the plait	30	—	30	—
A large ship's mast from 65 to 70 feet, 16–18 inches in the plait	25	—	25	—
A large ship's mast of 60 feet, 12–16 inches in the plait	20	—	16	—
An oak beam 45–50 feet long, 14–15 inches in the plait	7	—	5	—
An oak beam 40–45 feet long, 11–13 inches in the plait	5	—	3	—
A beech to the ship's keel is sold cubic by the cubic foot	—	4	—	3 or 6

SOURCE: APS, 65/34/0/-/5, F. 30–31. *FORSTORDNUNG FÜR POMMERN VON ANWEISUNG, FÄLLEN, VERKAUF, TAXE, UND LIEFERUNG, DES HOLZES, AUCH EINMIETHEN IN DER HEIDE*

from the ground; it is best to put a cloth beneath the tree and shake the acorns onto it. The same applies to storage, where the seeds cannot lie on the ground and should be stored somewhere that is neither too dry nor too humid. The author of the manual, Schulenburg, emphasised that oak grows in many soils, often in bad ones, but that these affect growing and the wood is of poor quality. If the ground is very poor, the tree will be small, will not produce acorns, and will never have a good straight trunk. If the soil is too fertile, it will grow beautifully but its wood will not have the necessary softness, firmness, and durability, which are important when oak is used in the construction industry. Therefore, oak trees tend to yield the best wood in areas where the soil is neither too fertile nor too poor. When the young oaks have reached a height of 10–12 feet and a thickness of 1–2 inches, all nearby trees of other species must be cut down as the young oaks must have space to grow. Oaks that grow too densely will be watched and the worst ones will be carefully dug up and planted elsewhere, 8–12 feet apart. Otherwise, the density of oaks and other trees will affect their growth and the overshadowed ones will not grow.[29]

All regulations issued in the second half of the 18th century in Silesia and Pomerania are examples of the progressive thinking of royal administration officials, who realised the importance of organising forest management, which was an important branch of the local economy. Thanks to these measures and the support of foresters and fiscal and military chambers, it was possible to take control of tree felling and introduce taxes on various types of wood, depending on their destination. A tree-planting system was introduced, which was to increase afforestation and thus lead to further use of wood as a strategic building material used in civil and industrial construction and the shipbuilding industry. In contrast to the 1748 *Ordenanza*, which was issued to guarantee the forest resources of the Spanish Navy, the control and care of afforestation in Prussia was aimed at ensuring the continuity of wood supplies for various branches of the local and national economy, not only for shipbuilding. The common feature of the Spanish and Prussian edicts was the priority given to either economic or political interests (building warships), which took precedence over mere concerns about forest conservation or deforestation.

3   Szczecin and the Spanish Connection during the Second Half of the 18th Century

In the Early Modern Age, Szczecin/Stettin maintained its medieval commercial links with Hanseatic ports as well as Denmark, Sweden, the Netherlands, eastern England, and northern France. In the first half of the 18th century, both the merchants and the city council, as well as the Prussian King Frederick William I, were looking for opportunities to trade with other crowns, opening new shipping routes to Russia and East Prussia.[30]

---

29   Ibid., fs. 42–45.

30   Henryk Lesiński, "Rozwój handlu morskiego Szczecina XVI–XVIII", in Hubert Bronk (ed.), *Estuarium Odry i Zatoka Pomorska w rozwoju społeczno-gospodarczym Polski* (Szczecin: Uniwersytet Szczeciński, 1990); Józef Stanielewicz, "Zarys rozwoju portu i handlu morskiego Szczecina od XVI do XVIII wieku", in Paweł Bartnik and Kazimierz Kozłowski (eds.), *Pomorze Zachodnie w tysiącleciu* (Szczecin: Uniwersytet Szczeciński, 2000), 121–128.

The next step in the development of Szczecin/Stettin and Prussian commercial networks was the countries of southern Europe, such as Portugal and Spain, which were seen as potential sources of wine, salt, colonial products, and, naturally, silver and gold. In 1719, Frederick William I encouraged Szczecin/Stettin based merchants to conduct direct transactions in Spain as a way to access the silver from the Spanish overseas colonies in America. The merchants from Szczecin/Stettin, however, were not particularly interested in direct trade with Spain due to the strong competition posed by the Dutch, British, and Scandinavian merchant fleets, which traditionally acted as middlemen in the trade between the Baltic and the Iberian crowns. Moreover, merchants from Szczecin/Stettin argued that their ships were too small to trade with Spain and that they did not have enough capital to enter the Iberian market permanently. One final issue was the lack of understanding of the goods demanded by Spain, as theretofore this trade had been based only on the sale of linen and cloth. The issue of trade with Spain reappeared virtually every decade during the 18th century. As a result, a Prussian consulate was established in Cadiz in 1748 to assist merchants from Szczecin/Stettin and other harbours.[31] In 1751, a trade college was created in Szczecin/Stettin, which included influential merchants from the city. The institution aimed to develop the harbour and set the guidelines for cooperation with other countries. In 1755, the newly appointed board of the college, which included Andreas Barthold,[32] Friedrich Schröder, Isaak Salingre, and others, tackled the issue of Szczecin's trade with Spain and Portugal.[33] Breakthroughs were made in the 1770s and 1780s, when new consulates were established in Alicante (1770), Valencia (1770), and Barcelona (1777), and two new companies were set up. The first was the Trading Company to Cadiz (Ger. *Handlungs Compagnie nach Cadix*) created in 1771 with the active participation of Silesian merchants who had previously worked with Spanish contractors in the flax and linen sectors;[34] the second, the Prussian Salt Company (Ger. *Saltzhandlungs-Compagnie*), was established in 1782 and had

---

31 The first Prussian consul in Cadiz was Silvester de Livorn.

32 From the decade of 1740 onwards, this merchant was active in the wood trade with Swedish contractors, selling them boards and beams, for instance in 1742, when he sent 800 pieces of 3–4-inch boards, 400 pieces of 5–8-inch boards, and 60 oak beams from Szczecin/Stettin. Radosław Gaziński, *Handel morski Szczecina w latach 1720–1805* (Szczecin: Wydawnictwo Naukowe Uniwersytetu Szczecińskiego, 2000), 284.

33 Gaziński, *Handel morski Szczecina w latach 1720–1805*, 276–277.

34 APW, Archiwum Giełdy Kupieckiej, exp. 82/304/0/-/347, fs. 18–52 "Acta von Etablierung einer Handlungs Copagnie und eines Handlungs Hausses Cadix Bedarf eines unmittelbahren Commerci mit Spanien und denen sich dabey zu Interessierenden Schlesien Kaufleuthen".

representatives in Madrid and Cadiz. Szczecin/Stettin merchants and their ships participated in both companies.[35]

Going back to the issue of timber, as previously noted, firewood, construction timber, barrel timber, and ship timber were shipped from the port of Szczecin/Stettin. The main destinations were Denmark, the Netherlands, Britain, and, to a lesser extent, France. Spain was only a marginal player in the timber trade. However, the record is clear that Szczecin/Stettin was heavily involved in the trade of forest resources. According to data from maritime traffic in the Sund, in the second half of the 18th century, a total of 49 ships left Szczecin/Stettin with timber for the Spanish shipbuilding industry.[36] Partial confirmation of these data can be found in the maritime traffic registers kept by the publishers of the *Wochentlich-Stettinische Fragund Anzeigungs-Nachrichten*, which in the 1750s record that 48 ships were sent from Szczecin/Stettin to three ports in Spain: Cadiz (20 vessels), Malaga (20 vessels), and San Sebastian (eight vessels); 31 of these ships hauled timber for barrels, and timber for shipbuilding was registered in ten instances. The other ships were loaded with grain or mixed cargoes.[37]

4   Filip Chone's Visit to Szczecin and Proposals for the Supply of Prussian Timber to the Spanish Navy in the 1770s

Chapter 1 mentioned that, in 1768, Felipe Chone made his first trip to the Baltic to inspect the properties of local hemp and wood. In his following trip, in 1775, the *asentista* visited several harbours in the region, including Szczecin/Stettin. Chone stayed in the city in August, and at the end of his visit, he wrote a report about the wood available there. His report, divided into three sections, describes the qualities and dimensions of oak boards, oak beams, and oak curves (see Tables 23 and 24).[38]

Concerning curves and *ligazones*, the first-class pieces of oak wood were offered for 12 groschen per cubic foot and the second-class ones for 8 groschen per cubic foot. Chone stressed that the woods were of good quality and were always evaluated and examined by an inspector appointed by the magistrate. When carrying out the transaction, the wood was put in the water after taking

35   Gaziński, *Handel morski Szczecina w latach 1720–1805*, 240 and 253.

36   Rafał Reichert "Direct Supplies of Timbers from the Southern Baltic Region", *Studia Maritima* 33 (2020), 137–138; ten ships in the 1750s; seven in the 1760s; 20 in the 1770s; ten vessels in the 1780s; and two in the 1790s.

37   Emil Chróściak "Szczecin's Maritime Timber Trade and Deliveries to Spain", 158.

38   AGS, SMA, Asientos, leg. 624, Demostración sobre madera de roble.

TABLE 23   Dimensions and prices of oak boards in Szczecin/Stettin as described by Chone (1775)

Board thickness	Feet long	Price per cubic foot
3-inch thick boards	30	8 groschen[a]
4-inch thick boards	32–36	10 groschen
5, 6, 7, 8 and 9-inch thick boards	32 and more	12 groschen

a   Eng. pennies.
SOURCE: AGS, SMA, ASIENTOS, LEG. 624, DEMOSTRACIÓN SOBRE MADERA DE ROBLE

TABLE 24   Dimensions and prices of oak beams in Szczecin/Stettin as described by Chone (1775)

Beam class	Feet long	Inches in square	Price per cubic foot
First class	24	12–16	9 groschen
Second class	23–18	12–16	8 groschen
Third class	17–12	12–16	7 groschen

SOURCE: AGS, SMA, ASIENTOS, LEG. 624, DEMOSTRACIÓN SOBRE MADERA DE ROBLE

its measurements. In this operation, the buyer or his representative could participate. Chone also added important information on measurement systems; Rhenish and Prussian feet were larger than the Spanish (91⅓ were equal to 100 Castilian feet). In addition, the *asentista*, whose calculations were in bullion reales, reckoned that the lion's share of the cost went to freighting and insurance. Finally, Chone also pointed out that during his trips to the Baltic, he was not only looking for timber and hemp for the Spanish Navy but also other goods. "In the ports of the Baltic, I have surveyed other resources that are also necessary for the Royal Navy, mainly salted meats, canvas, tin sheets."[39] This information was also in the notes that he shared with officials at the Secretary of the Navy.[40]

---

39   "En los puertos del Báltico me ha dedicado al conocimiento de otros objetos que son necesarios al consumo de la Real Armada, y principalmente al de las carnes saladas, lienzos vitres, hojas de lata".
40   AGS, SMA, Asientos, leg. 624, En cumplimiento del venerado precepto de V. Ex. y de mi anhelo de hacerme útil al Real servicio.

In addition to Chone's activities, the Prussian side also began making over-tures to supply the Spanish Navy with timber for shipbuilding. In December 1776, the Prussian plenipotentiary minister in Paris, Neudi, presented a project for the supply of wood from Prussia for the construction of ships in Spain to the Spanish ambassador, Count of Aranda, including timber from Saxony, Branden-burg, and the upper banks of the Elbe River. As most of these forests belonged to the king, the general administration of the wood trade for the king of Prussia had been set up in Hamburg, and from 1776 this had supplied the French Navy with wood for shipbuilding. The destination harbour for this wood was Brest. In the offer he presented to Spain, Neudi committed Prussian contractors to serve different carved parts such as tables, curves, and *genoles*. In April 1777, the pro-posal was rejected by the Count of Aranda, who in his letter to Minister Grimaldi explained that the Prussians offered three different classes of wood at different prices, but that the offer did not improve on Spanish *asientos* already in place in the Baltic. Also, Neudi did not explain how this wood was to be shipped.[41]

At the same time, a merchant from Szczecin/Stettin, Johann Gotte Hering,[42] also made a proposal to provide oak and pine wood from Silesia and Poland to the Spanish Navy. On 13 May 1777, during a visit to Spain, Hering offered to bring to El Ferrol three shipments of straight wood, curves, *genoles*, and oak planks from 3 to 9 inches, at different prices, ranging from 96 to 144 reales de vellón per cubic cubit for each carved piece. In his letter, Hering explained that he had experience in the wood trade, supplying Denmark and Sweden, which were crowns that, despite having their own forests, preferred to buy oak wood from Silesia and Poland, which they regarded as being of better quality. The merchant also offered to build war frigates in Szczecin/Stettin to the specifica-tions of the Spanish Royal Navy.[43]

In fact, after the end of the Seven Years' War in 1763, several frigates of between 30 and 50 guns were built at the Szczecin shipyard for Dutch and French contractors, as well as for King Frederick II, who later sold them to other monarchs. For example, in 1770 the Prussian king requested the con-struction of a large frigate at the Szczecin shipyard, which he later sold to the French Navy, which named it *Duc de Bievre*. However, the peak of shipbuild-ing in Szczecin/Stettin was yet to come: between 1781 and 1795, as many as 144 ships of various types were built at the shipyard.[44]

41   Ibid., Tratado de surtimiento de maderas de construcción de la Sajonia.

42   The merchant had contact with Felipe Chone during his visit to Szczecin in August 1775 and for this reason he presented the proposal for the contract for the supply of wood to the Spanish Navy. Ibid., Demostración sobre madera de roble.

43   Ibid., Don Juan de Hering, vecino de Stetin, en Prusia.

44   Gaziński, *Handel morski Szczecina w latach 1720–1805*, 175–176.

5   Timber and Its Trade in Gdańsk During the Second Half of the 18th Century

From the Middle Ages, Gdańsk/Danzig was an important commercial hub in the south Baltic, notably as a source of grain, potash, wood, linen, flax, and hemp for much of Western Europe. The port underwent significant commercial growth in the 16th and 17th centuries,[45] but also in the 18th century, when Gdańsk/Danzig remained a major harbour, as shown by the shipping registers in the Sund[46] and the local newspaper *Danziger Nachrichten* (1765–1795),[47] which published weekly freight data. The period between the end of the Seven Years' War (1763) and the outbreak of the American War of Independence (1775) was particularly profitable for the harbour (Table 25). The city's obvious decline from 1773 onwards was related to the first partition of the Polish-Lithuanian Commonwealth, which was imposed a year earlier by Russia, Austria, and Prussia. Poland lost direct access to Gdańsk/Danzig through Prussia, and this was a major setback for the transport of goods from the interior of the Commonwealth down the Vistula River, especially with the creation of two new Prussian customs points in Nowy Port (Ger. *Nuefahrwasser*) and Fordon,

---

45   Maria Bogucka, *Gdańsk jako ośrodek produkcyjny w XIV–XVII wieku* (Warsaw: Wydawnictwo Naukowe PWN, 1962); Czesław Biernat, *Statystyka obrotu towarowego Gdańska w latach 1651–1815* (Warsaw: PWN, 1962); Stanisław Gierszewski, *Statystyka żeglugi Gdańska w latach 1670–1815* (Warsaw: PAN, 1963); Maria Bogucka, *Handel zagraniczny Gdańska w pierwszej połowie XVII wieku* (Warsaw – Krakow: Ossolineum – PAN, 1970); Andrzej Groth, *Rozwój floty i żeglugi gdańskiej 1660–1700* (Gdańsk: Zakład Narodowy Ossolińskich, 1974); Artur Attman, *The Russian and Polish Markets in International Trade, 1500–1650* (Göteborg: Institute of Economic History of Gothenburg University, 1973); W. Heeres and L. Noordegraaf, *From Dunkirk to Dantzig: Shipping and Trade in the North Sea and the Baltic, 1350–1850* (Hilversum: Verloren, 1988).

46   According to records of direct trade from the Sund toll, from 1700 to 1783, Gdańsk/Danzig sent more ships to Spain (551 vessels) than any other South Baltic port. The total number of ships sailing out of the southern Baltic was 1,099, which means that Gdańsk/Danzig accounted for 50.14 per cent of the total, significantly ahead of Szczecin/Stettin (200 vessels, 18.20 per cent), Memel (113 vessels, 10.28 per cent), and Königsberg (94 vessels, 8.55 per cent). Rafal Reichert, "El comercio directo de maderas", 141–142.

47   In the years 1741–1757, the newspaper came out under various names: *Danziger Erfahrungen*; *Danziger Nachrichten, Erfahrungen und Erläuterungen allerley nützlicher Dinge und Seltenheite* and *Gemeinnützige Danziger Anzeigen, Erfahrungen und Erläuterungen alleringe alleringe Seltenheiten*. From 1758 to 1795 the newspaper used the name *Wöchentliche Danziger Anzeigen und dienliche Nachrichten*. Here I use the simplified form *Danziger Nachrichten*.

which led to a diminution of trade (and the related shipping activity) between Gdańsk/Danzig and the Kingdom of Poland.[48]

In the following years, Gdańsk's maritime traffic decreased, reaching its lowest level in the years 1780–1782 (see Table 25), when under 500 vessels left the port every year on average. The figures indicate that in the period 1775–1784, that is, during the American War of Independence and its immediate aftermath, an average of 582 ships per year passed through the harbour. Traffic continued growing in 1784–1786, peaking in 1786 with 1,011 ships (see Table 25). This was the last time in the period 1787–1795 that the number of vessels docking at Gdańsk/Danzig exceeded 1,000 in a single year. Afterwards, the average dropped to 621 ships.

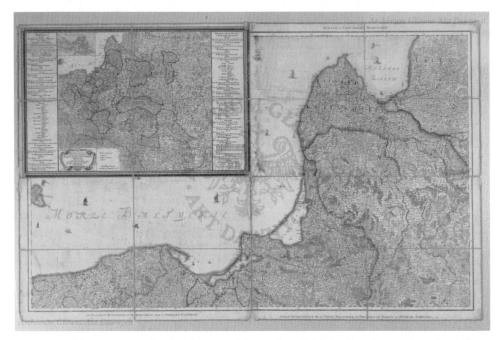

FIGURE 6   18th-century map showing the coast of the southern Baltic Sea from Pomerania to the Gulf of Riga
SOURCE: ARCHIWUM GŁÓWNE AKT DAWNYCH, 403 ZBIÓR GEOGRAFICZNY STANISŁAWA AUGUSTA, SYGNATURA 43, KARTA 33

---

48   Edmund Cieślak, "The Influence of the First Partition of Poland on the Overseas Trade of Gdańska", in Heeres and Noordegraaf, *From Dunkirk to Dantzig*, 203–215; Czesław Biernat, *Statystyka obrotu towarowego Gdańska w latach 1651–1815* (Warsaw: PWN, 1962), 22; Szymon Kazusek, *Spław wiślany w drugiej połowie XVIII wieku (do 1772 roku)* part 2 "Statystyka spławu wiślanego" (Kielce: Wydawnictwo Uniwersytetu Jana Kochanowskiego, 2016).

TABLE 25   Shipping in the port of Gdańsk/Danzig according to *Danzinger Nachrichten* (1765–1795)

Year	Incoming ships	Departing ships
1765	1,293	1,273
1766	1,128	1,129
1767	1,127	1,112
1768	1,151	1,177
1769	1,037	1,024
1770	1,250	1,240
1771	870	874
1772	1,021	1,031
1773	785	757
1774	682	684
1775	616	622
1776	602	604
1777	652	615
1778	548	532
1779	537	536
1780	446	456
1781	502	489
1782	446	449
1783	694	683
1784	839	830
1785	844	857
1786	1,014	1,011
1787	658	658
1788	409	419
1789	518	511
1790	492	503
1791	600	597
1792	653	655
1793	761	833
1794	830	814
1795	580	598

SOURCE: POLSKA AKADEMIA NAUK BIBLIOTEKA GDAŃSKA (PANBG), *DANZINGER NACHRICHTEN* EXP. X3518, FROM 1765–1795

TABLE 26   Number of ships loaded with timber dispatched to other European countries
             from Gdańsk/Danzig according to departure registers from *Danzinger*
             Nachrichten (1775–1784)

Year	Spain	France	Britain	Holland	Denmark	Other[a]	Total	% of all maritime traffic
1775	15	21	50	44	29	5	164	26.37
1776	9	29	93	14	25	2	172	28.48
1777	21	23	101	3	27	2	177	28.78
1778	34	8	96	9	15	3	165	31.02
1779	31	3	81	13	28	5	161	30.04
1780	2	14	79	11	18	5	129	28.29
1781	—	23	110	4	43	2	182	37.22
1782	23	52	114	2	75	12	278	62.33
1783	9	17	70	15	36	6	153	22.40
1784	8	44	105	11	28	8	204	24.58
1785	3	34	88	14	8	3	150	17.51
Total	155	268	987	140	332	53	1,935	29.01

a   Other ports and countries to which Gdańsk/Danzig timber was shipped: Memel, Lubeck,
Hamburg, Bremen, Oostende, Porto, Livonia, Sweden, Norway, Ireland, and Portugal.
SOURCE: PANBG, *DANZINGER NACHRICHTEN* EXP. X3518, 1775–1785

The analysis of the maritime traffic recorded by the *Danzinger Nachrichten*
shows that the largest number of ships loaded with timber went to Britain,
which accounted for more than half of these cargoes.[49] Ships loaded with timber arrived from the southern Baltic to Spain once every two or three weeks.
However, convoys were also organised, with three or four ships sailing out at
the same time. This is shown, for example, by issue 44 of the newspaper, dated
7 November 1778,[50] and issue 22, dated 5 June 1779,[51] which reported different

---

49   The registers only contain a basic description like "holz" (wood) and "holzwaar" (wood-
       ware) which makes the exact purpose of the wood uncertain. However, it can be assumed
       that it was wood for naval construction because the registers of the *Danzinger Nach-*
       richten also distinguish wood for barrels and firewood.
50   PANBG, *Danzinger Nachrichten* exp. X3518, issue 44, 7 November 1778, 523.
51   Ibid., issue 22, 5 June 1779, 263.

batches of four ships leaving Gdańsk/Danzig for Spain. An analysis of the captains' names shows that 80 per cent of them were Dutch, which suggests that Spain continued giving preference to their traditional Dutch freighters. More generally, it is interesting to note that in the years 1778 and 1779, when the shipment of Baltic wood to Spain from Gdańsk/Danzig peaked, Spain was the second largest customer, behind only Britain (see Table 26). In the period 1775–1785, Britain was always the main destination of timber from Gdańsk/Danzig. This confirms the Count of Aranda's impressions during his visit to Gdańsk/Danzig: based on information provided by the city council, he calculated that 90 vessels had departed to Great Britain loaded with timber,[52] which accounted for 53.89 per cent of the total number of ships setting sail from Gdańsk/Danzig with wood for shipbuilding.

The British predominance is also attested in other Baltic ports during the second half of the 18th century. For instance, the British were the main recipients of Swedish iron in the period 1760–1780; in 1759, London and Hull received 43.4 per cent of all Swedish iron, 34.7 per cent in 1770, and 27.2 per cent in 1780. Britain never lost its position as the leading importer of iron. From the 16th century, Sweden was the main supplier of iron in Europe, and during the second half of the 18th century almost 90 per cent of its maritime trade involved this metal. The same situation is attested for Norwegian timber, the main destinations of which were London, Hull, and Liverpool; in 1755, for instance, 13,500 deals by the Hundred were sent to Great Britain and 12,000 to other European destinations; ten years later, in 1765, 14,000 deals by the Hundred were sent to Britain and 13,000 to other ports; and, in 1780, 12,500 deals by the Hundred went to Britain and 10,000 to other destinations. Britain was also the main customer for Norwegian medium and small masts in the second half of the 18th century.[53]

The British predominance in the Baltic was also due to the size of the British merchant fleet, which in 1786 counted nearly 10,000 ships globally; the second largest fleet, the French, only had 5,268 ships, followed by the Danish – Norwegian (3,601), the Dutch (1,871), the Swedish (1,224), and the Spanish (1,202). This state of affairs allowed Britain, despite the armed conflicts in which it was involved, to hold a firm hegemonic position in most European markets,

---

52   AHN, Estado, leg. 4758, exp. 3. Noticias concernientes al comercio de la ciudad de Danzig. Año de 1761.

53   H.S.K. Kent, "The Anglo-Norwegian Timber Trade in the Eighteenth Century", *Economic History Review* 8, no. 1 (1955), 64–71; Ragnhild Hutchison, "The Norwegian and Baltic Timber Trade to Britain 1780–1835 and Its Interconnections", *Scandinavian Journal of History* 37, no. 5 (2012), 578–596.

including the trade in timber for shipbuilding, a material that was not only key for the Royal Navy but also for the huge British merchant fleet.[54]

In addition to the traditional river route down the Vistula, wood was also brought to Gdańsk/Danzig by ships from Eastern Pomerania, as recorded by the sea traffic registers from the *Danzinger Nachrichten*. For example, issue 41, dated 14 October 1775, records the arrival of two timber-loaded ships, commanded by Friedrich Reuzte and Captain Jan Schonbect, in the port.[55] These records and others—e.g. issue 31, dated 6 August 1785[56]—confirm that Gdańsk/Danzig also acted as an *entrepôt* for timber coming from the Königsberg, Pillau, and Memel regions. Further confirmation is found in the contract suggested by Gabriel Gotlieb Schultz to Ambassador Conde de Aranda, who met this influential trader during his visit to Gdańsk/Danzig in August 1761. The proposal involved bringing timber from Riga and other East Prussian ports to Gdańsk/Danzig and forwarding them to the departments of the Spanish Navy. Schultz explained that Gdańsk/Danzig rarely received 12.5 palms thick masts, adding that the logs sailing down the Vistula to Gdańsk/Danzig were not of as good quality as the kind harvested in Riga, Memel, and Königsberg, for which reason the *asentistas* were bringing pines masts from Riga.[57]

Data for the trade in timber coming from the hinterland of Gdańsk and floated down the Vistula show that the most desirable timber material was beams, and logs between 4 and 5 fathoms long and 3 and 4 inches thick (Table 27). Probably, this demonstrates that most of the timber harvested in the forests of the Kingdom of Poland was used to make planks, which were later mostly used in planking and dead works in European shipyards.

The shipment of large amounts of wood was possible through the convenient connection to Gdańsk's hinterland provided by the Motława and Vistula rivers. From the Middle Ages, the Vistula and the Bug, its most important right-bank tributary, were strategic links between Gdańsk/Danzig and the Polish interior regions. In the 16th and 17th centuries, the Vistula witnessed a boom in inland navigation, linking the surrounding area of Cracow with important commercial cities such as Sandomierz, Kazimierz Dolny, Warsaw, Płock, Toruń, Bydgoszcz, Grudziądz, and, finally, Gdańsk/Danzig. The river was used to float grain, wood, potash, tar, wax, honey, flax, hemp, iron, lead, leather, meat, cloth, and linen.

---

54   Hans Johansen, "Scandinavian Shipping in the Late Eighteenth Century in a European Perspective", *Economic History Review* 45, no. 3 (1992), 482.

55   PANBG, *Danzinger Nachrichten* exp. X3518, issue 41, 14 October 1779, 606.

56   Ibid., issue 31, 6 August 1785, 387. Captain George Bobl brought timber from the ports of East Pomerania.

57   AGS, SMA, Asientos, leg. 616. "El Conde de Aranda dice que habiendo estado en Dantzick, le propusieron porción de Arboladuras".

TABLE 27 Oak beams and logs floated down the Vistula River (1775–1785)

Inches	Fathoms	Pieces											Total
		1775	1776	1777	1778	1779	1780	1781	1782	1783	1784	1785	
1½	3	27	50	27	119	54	148	130	13	10	51	10	639
	4	23	57	67	37	160	702	279	189	110	29	27	1,680
	5	—	6	35	11	41	84	82	28	24	15	1	327
2	3	1,840	2,028	1,051	3,567	3,615	2,063	1,190	886	1,010	2,110	971	20,331
	4	1,918	2,468	1,574	2,431	4,105	4,751	3,740	2,049	2,127	1,876	989	28,028
	5	432	908	523	848	1,995	1,292	1,222	498	501	494	205	8,918
	6	24	136	81	107	193	186	129	53	49	67	31	1,056
2½	4	877	1,240	1,810	2,694	5,254	3,777	1,748	2,517	2,802	3,619	2,236	28,574
	5	377	566	531	2,015	2,930	1,471	1,197	1,233	1,069	1,950	1,021	14,360
	6	22	91	96	188	322	336	409	208	307	564	314	2,857
	7	5	8	1	22	55	53	51	30	42	83	28	378
3	4	4,288	2,355	1,994	6,347	7,667	8,910	10,474	9,774	11,360	11,855	5,742	80,766
	5	3,970	3,218	3,498	7,522	6,366	8,515	8,493	9,046	8,039	9,445	6,777	74,859
	6	1,102	1,239	1,014	2,002	2,750	2,285	2,898	3,243	2,817	4,430	2,453	26,233
	7	151	170	157	325	373	380	473	602	519	961	619	4,730
	8	15	8	1	26	31	75	60	74	77	140	82	589
4	5	3,064	1,786	2,774	2,494	5,807	6,062	7,348	8,103	10,416	10,251	4,657	62,762
	6	2,696	2,266	2,423	3,582	4,207	5,781	6,985	6,899	7,533	8,846	3,872	55,090
	7	974	1,046	839	1,257	1,808	2,133	2,542	2,471	2,639	3,708	1,491	20,908
	8	191	116	142	222	307	376	527	472	604	946	302	4,205
	9	39	19	7	28	28	51	65	80	126	166	30	639

TABLE 27 Oak beams and logs floated down the Vistula River (1775–1785) (cont.)

Inches	Fathoms	Pieces											Total
		1775	1776	1777	1778	1779	1780	1781	1782	1783	1784	1785	
5	5	17	56	29	134	153	40	51	210	229	207	45	1,171
	6	9	34	20	5	42	39	34	48	140	32	36	439
	7	2	10	—	5	12	4	5	20	27	18	1	98
6	5	32	145	65	69	72	61	106	186	258	18	4	1,016
	6	46	242	322	181	258	304	85	231	161	63	32	1,925
	7	14	78	35	48	41	35	8	50	119	5	8	441
	8	—	7	6	2	1	2	2	23	22	1	—	66
7	5	—	4	2	9	4	—	9	19	29	—	—	76
	6	3	15	—	22	23	23	45	20	49	3	—	203
	7	4	10	12	12	16	57	14	12	15	15	—	167
	8	1	1	—	3	3	4	—	1	3	—	—	14
8	5	4	4	2	2	—	—	73	16	32	1	—	134
	6	3	89	10	15	—	—	23	44	28	—	—	212
	7	5	21	1	5	2	4	4	14	7	—	—	63
	8	1	5	—	3	1	1	—	—	2	—	—	13
9	5	—	—	—	2	—	—	—	2	—	—	—	4
	6	—	1	1	5	—	—	19	1	1	—	—	28
	7	—	—	—	1	—	—	2	1	—	—	—	4
Total		22,176	20,503	19,150	36,367	48,696	50,005	50,522	49,366	53,303	61,922	31,995	444,005

SOURCE: CZESŁAW BIERNAT, *STATYSTYKA OBROTU TOWAROWEGO GDAŃSKA W LATACH 1651–1815* (WARSAW: PWN, 1962), 224–225

These commodities came from the regions of Wielkopolska, Kujawy, Mazowsze, Podlasie, and Małopolska, and even from the distant territories of Podole and Wołyń.[58] No data exist to establish the volume of timber sailed down the Vistula River with any certainty. The data available mainly comes from the royal accounts in Gdańsk/Danzig and from customs offices on the Vistula River, which were located, among other places, in Cracow, Sandomierz, Kazimierz, Puławy, Warsaw, Włocławek, Grudziądz, and Biała Góra. River custom posts were often leased to Jewish merchants or the nobility in exchange for a fixed fee.[59]

Concerning Crown-owned goods, we also have accounts of product turnover on the Narew and Vistula rivers for the period 1765–1767. The accounts concern timber harvested in the Białowieska Forest, because the ledgers were signed by A. Świętochowski, the commissar of Gdańsk Trade, who prepared them in the villages of Suchopol and Białowieża. The supervisor was Antoni Tyzenhauz,[60] the general economic manager and administrator of the wood trade in the Grand Duchy of Lithuania, under the jurisdiction of which these virgin forests were at the time. The timber felled in Białowieska was sent by the Narew River to the Bug, which flows into the Vistula, and from there the wood was directly floated to Gdańsk/Danzig.[61]

Another interesting description of the royal forests is provided by a survey, carried out by a forester called Biernacki, of the Kozienice Forest, located in the upper reaches of the Vistula River, between Kazimierz and Warsaw, in April 1793. The official sent his report to the Economic Committee of the Treasury and King Stanisław August Poniatowski. In the document, Biernacki distinguishes "barren and bare" from "good and dense" forest. He also divided the

---

58   Jan Małecki, *Związki handlowe miast polskich z Gdańskiem w XVI i pierwszej połowie XVII wieku* (Wrocław: Zakład Narodowy Ossolińskich, 1968), 7; Stanisław Gierszewski, *Wisła w dziejach Polski* (Gdańsk: Wydawnictwo Morskie, 1982), 22; Zbigniew Binerowski, "Transport wiślany w dawnej Rzeczypospolitej", in *Dolina Dolnej Wisły* (Wrocław: Wydawnictwo Polskiej Akademii Nauk Ossolineum, 1982), 283–297.

59   Czesław Biernat, *Statystyka obrotu towarowego Gdańska w latach 1651–1815* (Warsaw: PWN, 1962); Edward Stańczak, *Kamera saska za czasów Augusta III* (Warsaw: PWN, 1973), 52–57; Maria Bogucka and Henryk Samsonowicz, *Dzieje miast i mieszczaństwa w Polsce przedrozbiorowej* (Wrocław: Ossolineum, 1986), 198, 417–428; Józef Kus, "Z dziejów handlu Kazimierza Dolnego w XVII–XVIII wieku: instruktarze cła wodnego z 1616 i 1763 roku", *Rocznik Lubelski* 31–32 (1989–1990), 235; Szymon Kazusek, *Spław wiślany w drugiej połowie XVIII wieku (do 1772 roku)* part 2 "Statystyka spławu wiślanego" (Kielce: Wydawnictwo Uniwersytetu Jana Kochanowskiego, 2016).

60   Stanisław Kościałkowski, *Antoni Tyzenhauz, podskarbi nadworny litewski*, vol. 1 (London: Wydawnictwo Społeczności Akademickiej Uniwersytetu Stefana Batorego, 1970).

61   Archiwum Główne Akt Dawnych (AGAD), Archiwum Kameralne (AK), reference number III/262, microfilm A-44994. "Rachunek Handlowy Towarów Leśnych Spławu Gdańskiego".

TABLE 28   Extraction, storage, and trade of timber from the Białowieska Forest in 1765

	Stave [sixty]	Beams [pieces]	Planks [fathoms]	Logs [fathoms]
Timber harvested in 1765	261	234	—	—
Total timber stored in previous years	985	383	2,940	—
Timber floated and sold in Gdańsk/ Danzig	538	383	316	—

SOURCE: AGAD, UNIT AK, REFERENCE NUMBER III/262, MICROFILM A-449944. RACHUNEK HANDLOWY TOWARÓW LEŚNYCH SPŁAWU GDAŃSKIEGO

TABLE 29   Extraction, storage, and trade of timber from the Białowieska Forest in 1766

	Stave [sixty]	Beams [pieces]	Planks [fathoms]	Logs [fathoms]
Timber harvested in 1766	1,033	1,091	1,087	—
Total timber stored in previous years	1,480	1,091	3,711	—
Timber floated and sold in Gdańsk/Danzig	926	380	890	—

SOURCE: AGAD, UNIT AK, REFERENCE NUMBER III/264, MICROFILM A-449946. RACHUNEK HANDLOWY TOWARÓW LEŚNYCH SPŁAWU GDAŃSKIEGO

TABLE 30   Extraction, storage, and trade of timber from the Białowieska Forest in 1767

	Stave [sixty]	Beams [pieces]	Planks [fathoms]	Logs [fathoms]
Timber harvested in 1767	1,562	711	149	51
Total timber stored in previous years	2,116	1,422	2,970	51
Timber floated and sold in Gdańsk/Danzig	1,932	849	1,946	51

SOURCE: AGAD, UNIT AK, REFERENCE NUMBER III/263, MICROFILM A-449945. RACHUNEK HANDLOWY TOWARÓW LEŚNYCH SPŁAWU GDAŃSKIEGO

soils into different categories: sandy, dominated by shabby pines and birches; hard and good, populated by pines, birches, and oaks; and moist and fertile, inhabited by pines, birches, oaks, beeches, alders, firs, maples, ashes, lindens, white beeches, and other species. Biernacki points out that the lack of rational forest management had led to a scarcity of young pines, fir, and oak trees, and adds that trunk density did not allow for the easy growth of young trees. In addition, he points out that many handsome pines were being burned, which means that no profit could be made from them because they could not be floated to Gdańsk/Danzig or sent to a sawmill and were only suitable for fuel. This was caused by the locals' pernicious habit of setting the forest on fire.[62]

In the following sections of the report, the forester focused on tree felling. He pointed to uncontrolled cutting and lack of effective clearing in felling areas, which made it difficult for young trees to grow. He claims that felling in the vicinity of the town of Kozienice was so intense that there were no trees left that could be sent to Gdańsk/Danzig and Warsaw. In addition, he drew attention to the fact that good construction timber was located far west of the Vistula, making transport to the river very expensive and unprofitable for the royal economy. On the other hand, the shallows of the Radomka River, a tributary of the Vistula, were so covered in wood that in 1793 rafting had ground to a complete halt. The official mentions that, in 1791, contracts were signed for the felling and floating of beams and staves, notably a contract granted by the local royal administration to the Jew Leybus from Kozienice, who that year produced 140 beams, which were sent to Gdańsk/Danzig. Also in 1791, the merchant Dangel signed a contract to cut wood to be delivered to the royal sawmill in Bronki to make logs and boards.[63]

Biernacki also claims that the royal administrator had been trying to improve the situation in the Kozienice Forest for years, but that the foresters and rangers in his payroll did not want to fight the practice of burning or illegal logging. Similarly, they did not enforce logging contracts. The official's main conclusions were as follows:

1. Cutting was out of control, and no efforts were made to dispose of felled trees, although young ones were still planted, despite the abundance of trees.

2. Forests were not cleared of felled and blown-over trees, which posed no small obstacle to growing trees.

3. Due to recurrent fires, many of the trees were useless.

---

62 AGAD, unit AK, reference number III/260, Dopełniwszy obowiązek zalecony mi w objechaniu y zlustrowaniu Lasów JKM Kozienickich, 112–114.

63 Ibid., 115–116.

4. The clearing of large areas of the forest near Kozienice and near the village of Stanisławów for fields and meadows created a problem for the transport of wood to the Vistula by road.

5. The bad habit of constantly breaking up new soil prevented the forest from regenerating and left abandoned agricultural land barren.

6. Locals took livestock to graze in the forest, and this destroyed young trees, and adult trees lost their bark, leading to the release of resin and the trees drying out.

7. Young, healthy, and beautiful pines and oaks, from 30 to 40 years old, were cut down for fuel, charcoal, and fences because pines burned easily and oaks were even and suitable for garden piles.

8. Due to the earlier felling of trees to be sent to Warsaw and Gdańsk/ Danzig, the forest had thinned out, and violent winds broke trees more easily.

9. Some areas had become so overgrown with grass that the seeds falling from trees did not germinate or did so in untidy ground and decayed.

10. Most of the soil was poor, and trees required many years to grow.[64]

6   Some Information about Forestry Knowledge and Forest Management Policy in the Polish-Lithuanian Commonwealth during the Second Half of the 18th Century

It should be emphasised that the forests belonged to the Crown and the nobility. Often, influential aristocratic families leased royal land (Pol. *królewszczyzny*) for their own use. In the Grand Duchy of Lithuania, large royal estates were managed by such families as the Radziwiłł, the Sapieha, the Massalki, and the Branicki.[65] Royal forests, despite being managed directly by royal officials— the Grand Crown Huntsman or the Lithuanian Grand Huntsman—and their foresters, guards, and rangers, were not always adequately protected against the predatory activities of peasants and noblemen who illegally sourced them for wood or felled them to expand their agricultural estates.[66] An attempt to

---

64   Ibid., 117–118.

65   Otton Hedemann, *Dawne puszcze i wody* (Wilno: Księgarnia Św. Wojciecha, 1934); Antoni Mączak, *Klientela. Nieformalne systemy władzy w Polsce i Europie XVI–XVIII w.* (Warsaw: Państwowy Instytut Wydawniczy, 1994); Ewa Dubas-Urwanowicz and Jerzy Urwanowicz, *Magnateria Rzeczypospolitej w XVI–XVIII wieku* (Białystok: Wydawnictwo Uniwersytetu w Białymstoku, 2003).

66   Michał Kargul, "Administracja leśna w dobrach królewskich w świetle lustracji województwa pomorskiego z 1765 roku", *Acta Cassubiana* 10 (2008), 60; Grzegorz Buczyński,

reform the management of the royal forests of Niepołomicka, Kozienicka, Sandomierska, Tuchola (to 1772), Białowieska, Persztuńska, Przełomska, Nowodworska, Olicka, Kidulska, Stryjowska, and Bersztańska, located both in the Polish Crown and in the Grand Duchy of Lithuania, was undertaken in 1765–1780 by Antoni Tyzenhauz, an official that enjoyed the trust of King Stanisław August Poniatowski. Tyzenhauz separated forest areas from agricultural land and divided forests into plots to measure them and create maps with which to improve their administration and increase control over the management of forest resources. However, his reforms met with resistance, especially among the aristocrats who leased royal land in the Grand Duchy of Lithuania, who resented their forestry practices being culled, notably the Radziwiłłs, who reaped enormous profits from these forests but left them devastated.[67]

In any case, there were also powerful magnates who were interested in the protection of their forest resources, such as the Grand Chancellor of the Crown, Andrzej Zamoyski, who in 1775 issued a document entitled *Instruction on the Manner of Forest Management* for his estates in Krzeszów and Cieszanów.[68] In this document, he regulates felling policies and ordered the planting of new trees to replace felled ones. Zamoyski renewed the ordinance in 1785. It is interesting to note that the Zamoyskis were familiarised with Prussian forest legislation and that the Grand Chancellor invited German foresters to manage his woodland.[69]

Also in the second half of the 18th century, the Seym debated the need to protect the royal forests. However, a specific legal act was issued only in 1778, when, already after the first partition of the Polish-Lithuanian Commonwealth, King Stanisław August Poniatowski issued the *Universal about Relative to Forests and Conifer in the [Polish] Crown and in the Grand Duchy of Lithuania*,[70] confirming the ravaged state of the country's forests: "citizens show no moderation in managing the forests; encouraged by profit, they ravage them".[71]

"Podejście prawne do ochrony lasów w Polsce w ujęciu historycznym", *Kwartalnik Prawa Publicznego* 8 nos. 3–4 (2008), 11.

67 Otton Hedemann, *Dzieje puszczy Białowieskiej w Polsce przedrozbiorowej w okresie do 1798 roku* (Warsaw: Instytut Badań Lasów Państwowych, 1939), 48; Otton Hedemann, "Wrąb Radziwiłłowski", *Echa leśne* no. 38 (1935), 4.

68 Pol. *Instrukcja o sposobie prowadzenia lasu*. Krzysztof Okła, *O początkach regulacji leśnej w Puszczy Kozienickiej* (Kozienice: Nadleśnictwo Kozienice, 2019), 30.

69 Józef Broda, *Historia leśnictwa w Polsce* (Poznań: Wydawnictwo Akademii Rolniczej im. Augusta Cieszkowskiego, 2000), 48–49.

70 Pol. *Uniwersał Względem Borów y Lasów w Koronie y w Wielkim Xięstwie Litewskim*. Buczyński, "Podejście prawne do ochrony lasów w Polsce w ujęciu historycznym", 11.

71 Ibid.

This damage was mainly related to the increased demand for wood for house and ship construction as well as for the production of tar and potash, largely intended for export to England, the Netherlands, Denmark, France, and Spain.

7   Spanish Knowledge about Poland, Based on the Description of the Count of Aranda during His Trip to Gdańsk

The Count of Aranda made a diplomatic trip from Warsaw to Gdańsk/Danzig in August 1761, with the aim of reviving trade between Poland and Spain. During his journey down the Vistula, he wrote an interesting description of the river. He begins by noting that the sources of the Vistula or Weixel River are in Upper Silesia and the Principality of Teschen (Pol. Cieszyn). Then the river enters Poland and passes through Cracow, Sendomir (Pol. Sandomierz), Warsaw, and many other towns, especially after it enters Royal Prussia. From the Silesian border, it is already navigable for timber rafts and other purposes, so that, including all its meanders, it forms a 120-mile-long navigable waterway. It always runs smoothly, and there is not a sector in its course that cannot be passed at night as well as during the day when the flow is strong enough. It is, in fact, too wide, but there is no solution for this because the whole country is flat and when the water rises it brings so much sand that the bed keeps rising and the waters spread more widely; this could only be solved with a continuous dam, which would be very costly. From Cracow to Thorn (Pol. Toruń), in Prussia, both shores are covered by forest, which only thins out in the vicinity of Warsaw; in Prussia, the forests are less thick and farther from the water. Five miles below Warsaw, the Buch (Pol. Bug), which flows down from Lithuania and is also navigable for most of the year, empties into the Vistula.[72]

The Spanish ambassador noted that no other river in Europe would be more comfortable for navigation nor could be more useful should the inhabitants of its banks be less indolent and make more of the goods brought by ships. Throughout its course, the banks are so equal that the towpaths are no more than one or two yards above the river, so ferrying the boats could not be more comfortable for the horses. It is true that the river, particularly below Warsaw, has many islands; but all the channels on both banks have enough water, so only if the skipper is drunk, which they usually were, can boats run aground.

---

72   AHN, Estado, leg. 4758 exp. 3, Relación del comercio de la ciudad de Dantzig (en latín Gedanum) antiguamente ciudad hanseática; y hoy dependiente de la Polonia. Como una de las ciudades de la Prusia que conservaban aun su gobierno interior con Libertad.

The bottom of the river, which is frequently and carefully examined below Warsaw, has always been found to be firm and sandy and free from stones and mud. At Thorn, the first city of Prussia, the river passes under a wooden pilot bridge that usually suffers from the yearly floods, more because of the brush and sheets of ice that it brings than because of the speed of the water. Below Mewa, the Vistula River branches into two; the right arm goes through Marienburg (Pol. Malbork) towards Elbing (Pol. Elbląg) and ends in the Haff (Ger. Frisches Haff; Pol. Zalew Wiślany); this arm carries with it almost two-thirds of the water; the left arm runs towards Dirschaw (Pol. Tczew) and Gdańsk/Danzig. Three miles below Dirschaw and three miles above Gdańsk/Danzig, the left arm branches again into two: the much larger right arm takes a sharp turn to the right and runs into the Haff after four miles; the left arm, which by this point only carries from one-eighth to one-tenth of the overall flow, turns towards Gdańsk/Danzig, which is three miles from the point where the river branches out, and one mile past the city it enters the roadstead. Only the smaller branch that reaches Gdańsk/Danzig retains the name of Weixel or Vistula because the largest branch goes by the name of Nogat and the other by that of Haff. The Vistula is so small and is loaded so much that the greatest bottleneck for the boats is found a mile above Gdańsk/Danzig; boats often find no more than two feet of water and it becomes necessary either to lighten small boats or wait for the water to rise to reach the city; it can thus be said that the biggest obstacle is found in the most unlikely place.[73]

Finally, Aranda concludes that the largest branch of commerce in Gdańsk/Danzig is dealing with Polish goods and that the entire region lives off this trade. The Poles find many buyers for their grain, which is sold even before it has arrived, and have the certainty to find the same buyers again in the future. The people of Gdańsk/Danzig could not live without the Poles and, reciprocally, neither could the latter live without the former. The Poles carry out this trade by going downriver themselves, bringing top, medium, and inferior quality wheat; barley; ashes, which they call "*potasse*" and "*wedasse*"; construction timber and firewood for use in the city; hemp; linens; oils of different seeds to burn; wax; honey; Hungarian wines; coarse fabrics to pack meat for the supply of Gdańsk/Danzig, all of which are key for the Gdańsk/Danzig export trade.[74]

---

73   Ibid.
74   Ibid.

8   The Radziwiłł: Timber Felling and Trade with Königsberg, Memel,
       and Riga

This section presents the network created by the Radziwiłłs, one of the most influential families in the Polish-Lithuanian Commonwealth, for the extraction and commerce of timber in the forests under their stewardship.[75] The profits reaped from trade in various products constituted the basic, if not the main, source of income for this wealthy family.[76] The family's economic records present an extensive and well-organised system of production and trade, a veritable economic powerhouse in the eastern territories of the Polish-Lithuanian Commonwealth. The Radziwiłłs, like other aristocratic families, set up a complex system of trade, managed by qualified officials who controlled different branches of production and transport.[77] The Radziwiłłs, who exported most of their goods through the Baltic ports, had close contacts with the most important European merchants and commercial organisations in the territory of the Commonwealth.[78] Their complex manufacturing and commercial system provided the Radziwiłłs with financial stability and not only economic but also significant political leverage in the region at the turn of the 19th century.[79] It was during this period, after a short period of stagnation caused by the fall in demand for Polish grain (which was the Polish-Lithuanian Commonwealth's main export), that Polish wealthy families also found their place in the timber trade.[80] The main route used by Polish magnates to float the timber was the Vistula, but also its tributary, the Bug, which connected the fertile and

---

75   The sections 8, 9, 10, and 11 were elaborated and written by Ph.D. students Karolina
      Juszczyk and Daniel Prusaczyk (who were scholarship holders on the project "The Role
      of Wood Supplies from the Southern Baltic Region and the Viceroyalty of New Spain in
      the Development of the Spanish Seaborne Empire in the Eighteenth Century" funded
      by the National Science Centre), and their work was supervised by the author of this
      monograph.
76   Dagnosław Demski, "Naliboki i Puszcza Nalibocka –zarys dziejów i problematyki",
      Etnografia Polska 38, nos. 1–2 (1994), 68–69.
77   Darius Žiemelis, "The Structure and Scope of the Foreign Trade of the Polish-Lithuanian
      Commonwealth in the 16th to 18th Centuries: The Case of the Grand Duchy of Lithuania",
      Lithuanian Historical Studies 17 (2012), 111–113.
78   Maria Bogucka, "Żegluga bałtycka w XVII–XVIII wieku w świetle materiałów z archiwum
      w Amsterdamie", *Zapiski Historyczne* 84 no. 4 (2017), 126–127.
79   Anna Lesiak, "Kobiety z rodu Radziwiłłów w świetle inwentarzy i testamentów (XVI–XVIII
      w.)", in Urszula Augustyniak (ed.), *Administracja i życie codzienne w dobrach Radziwiłłów
      XVI–XVIII wieku* (Warsaw: Wydawnictwo DiG, 2009), 113. Bogucka, "Żegluga bałtycka w
      XVII–XVIII wieku w świetle materiałów z archiwum w Amsterdamie".
80   Bogucka, "Żegluga bałtycka w XVII–XVIII wieku w świetle materiałów z archiwum w
      Amsterdamie", 124–125.

forested south-eastern territories of Podole, Wołyń, and Podlasie with Elbląg and Gdańsk/Danzig. On the other hand, the aristocratic estates to the east and north-east of the Polish-Lithuanian Commonwealth depended on the Pregoła, Łyna/Alle, Neman, and Daugava rivers to bring their goods to the harbours of Königsberg, Memel, and Riga.[81]

The development of the British, Dutch, French, Spanish, Danish, and Swedish navies raised the demand for shipbuilding timber in the 18th century, leading merchants and wealthy aristocrats to create new centres of wood production around the forests in the Grand Duchy of Lithuania, in particular around the rivers Neman and Daugava. These changes, in turn, resulted in the development of a complex system of production, rafting, and sale of wood in aristocratic estates, and the intensification of contacts between Polish-Lithuanian nobles and merchants and their European counterparts.[82]

9   The Radziwiłłs' Land and Forest Estates

The first stage in the wood trade cycle was the preparation and selection of logging and wood-dressing areas. The analysis of this process is an important aspect in the study of the business model set up by the Radziwiłłs in the 18th century, and it first requires logging areas to be determined and the administration system to be understood. The records mention 33 toponyms that refer to forests or afforested estates. Almost all of them were located in modern Belarus, the few exceptions being in modern Lithuania, Russia, and Poland. These estates belong to a category known as "magnate estates". In the 18th century, there were over 20 "magnate estates" in the hands of over a dozen of the most important families in the Polish-Lithuanian Commonwealth. These lands, which were divided into *królewszczyzny* and *ordynacje*, were the largest and best-managed aristocratic estates in the country.[83] As aristocratic authority increased, some Crown-owned lands were leased to noble families, who also took over their administration.[84] The profits reaped from these leases were an important source of income for these magnate families. In the 18th century, the Radziwiłłs were in control of several royal lands, both in the

---

81   Hedemann, *Dawne puszcze i wody*, 37–38.
82   Ibid., 38.
83   Mariusz Kowalski, "Księstwa Rzeczpospolitej. Państwo magnackie jako region polityczny", in Grzegorz Węcławowicz (ed.), *Prace Geograficzne* (Warsaw: IGiPZ-PAN, 2013), 236–238.
84   Andrzej Jezierski and Cecylia Leszczyńska, *Historia gospodarcza Polski* (Warsaw: Wydawnictwo Key Text, 2010), 60–61.

Kingdom of Poland and in the Grand Duchy of Lithuania. Most of these leases were granted while Karol Stanisław was the head of the family in the 1680s and 1690s.[85] Karol was the nephew of the reigning king, Jan III Sobieski, who, in support of his family, handed him over the Crown's richest estates in Poland, including, among others, the Człuchów and Przemyśl *starostwo*,[86] and also a number of Lithuanian estates in the Krzyczew, Kamieniec, Brasław, Niżyn, and Vilnius districts. Despite their great economic importance, at least officially these lands were not the main sources of the timber sold by the Radziwiłłs.[87] According to forest laws, which began developing in the 16th century, royal forests were spared from tree logging, at least in relative terms. The regulations stated that, in the Crown lands, access to wood or brushwood was to be limited to local logging, organised by individual subjects in time of need and intended for construction and fuel.[88]

As a result, the activity of aristocratic families on leased land was also limited. It is now known, however, that the Radziwiłłs and other magnates clandestinely carried out mass felling also in Crown lands even if, in order to dodge the law, the records make no mention of royal forests as a source of timber. The toponyms and geographical names mentioned in the record suggest that the forests in the vicinity of Vilnius (north of the Jašiūnai forest, which belonged to the Radziwiłł family)[89] and in the vicinity of Newel and Połock (areas of the Newel, Uświat and Sieruck forests) were exploited intensively. At any rate, based on the record, it can be assumed that Crown lands provided only a small proportion of all the wood. They are also mentioned mainly in early documents from the 1740s and 1760s. The main source of wood for the Radziwiłłs, based on the number of mentions, were forest areas located within their family estates, called *ordynacje*, which constituted the main part of their "magnate estates".

Ordynacje[90] meant both the institution and the land on which it applied. This form of ownership is understood as a large noble estate with a characteristic

---

85   Bernadetta Manyś, "Przyczynek do badań nad rolą i funkcją lasów w XVIII-wiecznych dobrach radziwiłłowskich w Wielkim Księstwie Litewskim", *Studia i Materiały Ośrodka Kultury Leśnej* 15 (2016), 160–163.

86   Literally "eldership", an administrative unit established in the 14th century in the Polish Crown and later in the Polish-Lithuanian Commonwealth until the partitions of Poland in 1795.

87   Zbigniew Anusik and Andrzej Stroynowski, "Radziwiłłowie w epoce saskiej. Zarys dziejów politycznych i majątkowych", *Acta Universitatis Lodziensis* 33 (1989), 36.

88   Demski, "Naliboki i Puszcza Nalibocka – zarys dziejów i problematyki", 51–52.

89   These are the original names of forests in the Lithuanian and Polish languages.

90   Latin: *ordinatio*.

inheritance system, which ensured that it remained undivided and in the hands of the family. The laws of *ordynacje* of the Radziwiłłs, established in 1586, categorically forbade daughters and their descendants to inherit. In the absence of a male successor, *ordynacje* assets were to fall to the second line of the family (and therefore to the owner's brother or his male descendants). *Ordynacje* existed across the Polish-Lithuanian Commonwealth and had parallels in other European regions, such as entails or *mayorazgos*, among others.[91]

The Radziwiłłs' most important landed properties were the territories around the cities of Ołyka, Kleck, and Nieśwież. These *ordynacje*, apart from their important administrative role for the family, were also their main source of income. Particularly important was the area of Nieśwież, which included, among others, Naliboki with its glassworks and a huge forest complex. The exploitation of the woodland was extremely important for the Radziwiłłs' economy. The Naliboki forest, later divided into smaller forests, was their basic source not only of wood but also meat and other forest commodities throughout the 18th century.[92] At the same time, in the Nieśwież district, forest areas located to the south, such as the forests of Nieśwież, Arciuchowska, and Odcedzka, along with the primeval forests in the vicinity of Mikołajewszczyzna, were also exploited. In 1760, the property of the Radziwiłłs expanded to include Birże, Dubingiai, Biała, Słuck, and Kopyła. The forests located there were also used to harvest wood, especially after they were incorporated into the Radziwiłłs' estate. However, especially in the late 18th century, the Nieśwież district played a dominant role in the production and trade of wood (see Map 7).[93]

The earliest register of wood logging in the Radziwiłł estates dates to 1744–1747, specifically in the three primeval forests located in the Połock region, next to the northern Belarusian-Russian border.[94] As noted earlier, in lands that were not under the *ordynacje* regime, the royal law was in force, which was also

---

91 Tomasz Brodacki, "Uwarunkowania prawne ordynacji Radziwiłłowskiej i jej wojsk w Rzeczypospolitej Obojga Narodów", *Zeszyty Naukowe Uniwersytetu Przyrodniczo-Humanistycznego w Siedlcach* 109 (2016), 223–227.

92 Karl Friedrich Einchorn, *Stosunek xiążęgo domu Radziwiłłow do domów xiążęcych w Niemczech uważany ze stanowiska historycznego i pod względem praw niemieckich poltycznych i xiążęcych* (Warsaw: Księgarnia Aug. Emm. Glücksberga, 1843), 81–88.

93 Demski, "Naliboki i Puszcza Nalibocka—zarys dziejów i problematyki", 58; Manyś, "Przyczynek do badań nad rolą i funkcją lasów w XVIII-wiecznych dobrach radziwiłłowskich w Wielkim Księstwie Litewskim", 161–162.

94 AGAD, Warszawskie Archiwum Radziwiłłów (WAR), Kolekcja XX-Handel Rzeczny (1608–1892), exp. 32 and 80 for 1744–1746.

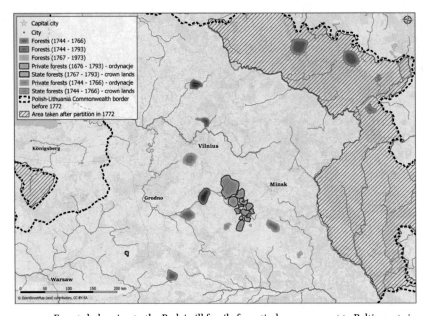

MAP 7   Forests belonging to the Radziwiłł family from timbers were sent to Baltic ports in
         Königsberg, Memel, and Riga
         SOURCE: LOOK4GIS-LUKASZ BRYLAK, KAROLINA JUSZCZYK, AND DANIEL PRUSACZYK
         BASED ON QGIS SOFTWARE

applied to forest management. Despite the restrictions, however, the Newel
and Siebież forests were used by the Radziwiłł family, especially as a source of
wood for masts, which were generally rafted to Riga. In 1746, there was also a
single mention of the rafting of logs from the Uświat *starostwo*.[95] Despite this
single mention, the harvesting of wood in Radziwiłł estates took place mainly
in the vicinity of Newel and Siebież. Even in 1761, when this locality frequently
features in the record, this includes only one mention of wood harvesting, in
the Sieruck Forest, which was situated in the Połock region, east of Vitebsk (see
Table 31).[96]

In the following years, the Lithuanian estates became the dominant sources
of wood for the Radziwiłłs.[97] At that time, the Dubingiai and Vyžuonos forests
played a significant role. They were the only places mentioned as sources of

---

95   Ibid., exp. 80 f. 13 (1746).
96   Ibid., exp. 32, fs. 1–4 (1761).
97   Ibid., exp. 32 fs. 18–24 and 112 f. 27 (1767).

TABLE 31   Radziwiłł wood production centres (1744–1766)

Forest	1744	1745	1746	1747	1761	1763	1765	1766
Amhowicka							X	
Dokudava							X	
Dubingiai						X		X
Hresk							X	
Lipiczańska							X	
Mogilno							X	
Newel	X	X	X	X	X			
Pogorzelce							X	
Ryhalska							X	
Siebież	X	X	X	X	X			
Sieruck						X		
Uświat				X				
Sverinovo							X	
Szichowicka							X	
Szutkowska							X	
Vyžuonos						X		X

Note: It should be emphasised that data from the years 1748–1760 are missing for this period. Therefore, they must be considered partial data, but sufficiently solid and reliable for the purposes of this monograph.
SOURCE: AGAD, WAR, KOLEKCJA XX-HANDEL RZECZNY EXP. 32, 80, 112 FOR THE PERIOD 1744–1766

wood in 1763,[98] 1766,[99] and 1767.[100] As noted, both forests are situated in the Duchy of Lithuania. The main locations for this branch of the Radziwiłł family were the towns of Dubingiai, located about 45 km north of Vilnius, and Vyžuonos (about 100 km north of the Lithuanian capital). This *ordynacje*, which also included forest complexes, played a crucial role for this magnate family. These forests are mentioned as production centres for different types of wood products, like logs, timbers, masts, and staves.[101] One key advantage

---

98   Ibid., exp. 112 fs. 23–24 (1763).
99   Ibid., exp. 32 fs. 10–16 and 112 f. 25 (1766).
100   Ibid., exp. 32; fs.18–24 and 112 f. 27 (1767).
101   Anusik and Stroynowski "Radziwiłłowie w epoce saskiej. Zarys dziejów politycznych i majątkowych", 44–45.

of the Dubingiai and Vyžuonos Forests was their location in the northeastern region of the Polish-Lithuanian Commonwealth, a relatively short distance away from two important Baltic ports: Königsberg and Memel.[102] During the period 1763–1767, these cities were the main hubs for the exportation of wood to other European countries. In addition, there is also information about the rafting of logs to Vilnius, which was probably the main storage centre for timber. The dominance of Lithuanian forests ended after 1767. In later years, there was only one mention of wood harvesting at the Dubingiai forest in 1778.[103]

The situation in 1765 is of great interest. It is the only year in the 1760s in which forests outside Lithuania are mentioned in the record. Two documents from this period yield information about the mass production of wood of various types in ten different forests:[104] Dokudava, Hresk, Lipiczańska, Mogilno, Pogorzelce, Ryhalska, Sverinovo, Amhowicka, Szichowicka, and Szutkowska. Most of them were part of the *ordynacja* of Nieśwież or were in its immediate vicinity. Dokudava and Lipiczańska are located in the western part of the *ordynacja*, where the Radziwiłłs used the wide local roads to transport their wood; the Mogilno, Pogorzelce, and Sverinovo forests, located in the southern part of the Nieśwież district, also played an important role as sources of wood for construction and also of potash; the Hresk forest was part of the former Novgorod province; only the Ryhalska forests were located outside the *ordynacja*, in the former jurisdiction of Polesie, to the south-west; this estate was famous for its vast forest complexes, used as sources of timber for masts, but also due to the proximity of the Białowieża, This estate was not, strictly speaking, owned by the Radziwiłłs, but was a Crown-owned estate. This sometimes led to friction between the magnate and royal officials, as the Radziwiłł loggers trespassed into the royal forests (see Map 7).[105]

From 1768, these forests were overshadowed by the Żahalska forest. This complex is mentioned time and time again in the years 1768–1780. It was located near the town of Żahale, in the jurisdiction of Novgorod,[106] under the administration of the Hlusk *starostwo* and part of the Radziwiłł estates from 1690, when they took over the city from the Czartoryski family. The region grew in importance owing to its extensive forest resources and good communication

---

102   Andrzej Groth, *Żegluga i handel morski Kłajpedy w latach 1664–1722* (Gdańsk: Wydawnictwo Uniwersytetu Gdańskiego, 1996), 6–14.

103   AGAD, WAR, Kolekcja xx-Handel Rzeczny, exp. 112, fs. 23–24 (1763); 112, f. 25 (1766); 112, f. 27 (1767) and exp. 32, fs. 10–16 (1778).

104   Ibid., exp. 112 fs. 9–15 (1765).

105   Ibid., exp. 32, f. 118 (1788).

106   Ibid., exp. 32, f. 30 (1768) and exp. 112, f. 35 (1768).

with Nieśwież, two factors that convinced the Radziwiłł family to develop a new wood production centre in the area.[107]

At the same time, the importance of the production centres at the Naliboki forest also increased. The complexes at the Swierzno forest and Włość Tulonka began to emerge as early as 1770 and 1774, at the same time as in Żahalska. Both areas were part of the *ordynacja* of Nieśwież, north of the town of Stołpce. Neither forest had been exploited on a large scale for wood production before, but these mentions indicate the growing role played by the region for the Radziwiłłs. In the second half of the 1770s, yet further wood production areas appeared. For example, between 1776 and 1778, the central part of the *ordynacja* of Nieśwież became increasingly prominent. There, the records especially mention the Orciuchy forest, located next to the southern borders of the Naliboki. This forest usually features in the record next to the Żahalska forest and was one of the two main sources of wood for the Radziwiłłs during the 1770s.[108]

Apart from these sites, the Naliboki complex, which occupied a large part of the *ordynacja* of Nieśwież, often features in 18th-century records. Currently, the Naliboki forest covers about 240,000 hectares, but during the rule of the Radziwiłłs, the forest expanded to over 350,000 hectares. This valuable woodland was exploited by the Radziwiłłs with great care. Tree felling and wood production moved seasonally across different areas of the forest in order to keep the woodland in balance.[109] This policy was probably behind the emergence of new logging areas, such as the Derevnoe and Tikhonova Sloboda forests.[110] The Khotova woodlands occupied the central sector of the Naliboki jurisdiction. This toponym is among the most frequently mentioned in the record in the period 1777–1779 and seems to have overtaken the Żahalska and Orciuchy forests in terms of production. The area is usually associated with the extensive production of beams and was regarded as the Radziwiłłs' main source of wood (see Table 32).[111]

In 1776 and 1778 there are also mentions of the Jašiūnai forest, located in Lithuania. This area, like Vyžuonos and Dubingiai, was owned by the Calvinist

---

107   Filip Sulimierski and Władysław Walewski, *Słownik geograficzny Królestwa Polskiego i innych krajów słowiańskich* vol. 7 (Warsaw: Druk Wieku, 1886), 78.

108   AGAD, WAR, Kolekcja XX-Handel Rzeczny, exp. 77, f. 48 (1776), exp. 32, fs. 82–83 (1777) and exp. 77, fs. 84–85 (1778).

109   Demski, "Naliboki i Puszcza Nalibocka –zarys dziejów i problematyki", 51–57.

110   AGAD, WAR, Kolekcja XX-Handel Rzeczny, exp. 32 fs. 114–120 (1788).

111   Ibid., exp. 32 fs. 82–83 (1777); exp. 77, fs. 30–31 (1777); exp. 32, fs. 92–99 (1778); exp. 77 fs. 84–85 (1778), and exp. 77 f. 133 (1779).

TABLE 32   Radziwiłł wood production centres (1767–1793)

Forest	1767	1768	1769	1770	1771	1772	1774	1776	1777	1778	1779	1780	1788	1789	1790	1792	1793
Orciuchy								X	X	X			X	X	X		X
Khotova									X	X	X						
Tytuvenai																X	
Derevnoe													X				
Dokudava													X				
Dubingiai	X									X							
Hościmska													X		X		
Jašiūnai								X		X							
Marchaczewszczyzna													X		X		
Mikołajewszczyzna																	X
Naliboki								X	X	X		X					
Parcewo								X									
Nieśwież															X		
Odceda													X	X			X
Pukhovshchina														X			
Tikhonova Sloboda													X				
Sverinovo																	X
Świerzno				X			X										
Vyžuonos	X																
Włość Tulonka							X										
Żahalska		X	X	X	X	X	X	X	X	X	X	X					

Note: It should be emphasised that data for the years 1773, 1775, 1781–1788, and 1791 are missing. Therefore, the data must be regarded as partial, but solid and reliable for the purposes of this monograph.

SOURCE: AGAD, WAR, KOLEKCJA XX-HANDEL RZECZNY EXP. 10, 31, 32, 77, 80, 112 FOR YEARS 1767–1793

line of the Radziwiłł family,[112] showing that wood sources were an important source of income for all family branches.[113] On the other hand, wood harvesting in the Parcewo forest, the area and town located inside the Białowieża forest, is interesting, because the name appears only once, in 1776. The exceptional nature of this operation may be confirmed by the fact that only the wood produced there was exported to Warsaw, and not, as was the case elsewhere, to Baltic ports. This register may also show the Radziwiłłs violating royal prohibitions, which forbade trees to be felled in Białowieża without permission.[114]

In 1788, for which the record is exceptionally rich, seven logging areas can be identified, mainly the Naliboki forest and some of its smaller sections, like the Derenvoe and Tikhonova Sloboda,[115] which were located to the south of the estate. In addition, the woodlands in the central sector of the *ordynacja* of Nieśwież, such as Marchaczewszczyzna[116] and Odceda[117] (see Table 32), are mentioned as beam production areas. Also, the vicinity of Orciuchy still played an important role as a source of timber. Interestingly, the Dokudava Forest reappears in the record for the first time in 23 years. This shows that this forest regained its former importance during this period and was one of the most productive logging areas in 1788.[118] The Hościmska forest is also cited;[119] the context of these mentions, and the fact that it is generally paired with the Marchaczewszczyzna forest, may suggest that it was one of the smaller forest areas located north of Nieśwież, at the centre of the *ordynacja*.

In the early 1790s, despite a clear decline in the importance of the wood trade, the Radziwiłłs still exploited the densely forested areas of Nieśwież, particularly in the Orciuchy,[120] Marchaczewszczyzna,[121] Mikołajewszczyzna,[122] and Sverinovo districts.[123] The economic records also mention the continued production of wood in the Naliboki forest. During this period, the production of timber in the Żahalska and Dokudava forests completely disappeared and there is no reference of the exploitation of woodland in the vicinity of Vilnius

---

112   Anusik and Stroynowski "Radziwiłłowie w epoce saskiej. Zarys dziejów politycznych i majątkowych", 44.
113   AGAD, WAR, Kolekcja XX-Handel Rzeczny, exp. 77 f. 48 (1776) and exp. 77 f. 85 (1778).
114   Ibid., exp. 31 f. 201 (1776).
115   Ibid., exp. 32 fs. 114–120 (1788).
116   Ibid., fs. 115, 120.
117   Ibid., fs. 115, 117.
118   Ibid., fs. 118–120.
119   Ibid., fs. 115–117, 120.
120   Ibid., exp. 32, fs. 155–158 (1790) and exp. 32, fs. 168–169 (1793).
121   Ibid., exp. 32, f. 156 (1790).
122   Ibid., exp. 32, f. 168 (1793).
123   Ibid.

(see Table 32). There is a single mention of the Tytuvenai forest,[124] which did not feature in the earlier records. This city belonged to the Radziwiłł family from 1706, when it was granted to Karol Stanisław Radziwiłł, after a long lawsuit. From that moment on, the house and *latifundia* of Tytuvenai came under the possession of the Nieśwież branch of the family. Due to the distance from the main *ordynacja*, the Radziwiłłs gave the tenants jurisdictional authority over Tytuvenai. Later, the forests on these estates may have become one of the most productive logging areas, taking advantage of their proximity to the port of Memel and their location, directly on the route between Memel and Vilnius.[125]

10 Production and Shipment of Wood from the Radziwiłł Estates in 1744–1793

The production of timber such as baulks, beams, logs, planks, masts, spikes, bowsprits, staves, and potash in the Radziwiłł estates was most often counted in pieces, and less frequently in other measurement units, such as fathoms or cubits. In the period 1744–1766, the production of a total of 23,026 different pieces of wood is attested (see Table 33).[126] The year 1765 is particularly interesting because all the wood came from the Nieśwież *ordynacja*. Despite this, the official registers mention the production of 10,679 pieces of wood, which accounts for over 45 per cent of the total for 1744–1766. This was driven by the orders from the European powers to restore their navies after the Seven Years' War. The Spanish Navy, for instance, made an order for at least 60 masts with a thickness of between 9¼ and 12 spans, which were to be brought from the Radziwiłł estates to Riga, and thence, together with beams and planks, forward to Cadiz.[127]

However, in the years 1767–1793, there was not such a large felling of trees at the Nieświż *ordynacja*, and the burden of logging was spread over other smaller forest areas, such as the Orcichy, Hościmska, Marchaczewszczyzna, and Odceda forests in the 1780s. In these years, a large number of beams and baulks were harvested in the Radziwiłł estates (30,736 pieces), much more than in the period 1744–1766. Additional output from their forests included 100

---

124   Ibid., exp. 10, fs. 16–17, 37 (1792).
125   Anusik and Stroynowski "Radziwiłłowie w epoce saskiej. Zarys dziejów politycznych i majątkowych", 46.
126   It should be remembered that these are partial data from this period.
127   AGS, SMA, Asientos, leg. 616 "Resumen de las 10 partidas de Arboladura gruesa que han de bajar al uerto de Riga en la primavera de 1766".

TABLE 33   Wood types (in pieces) produced in the Radziwiłł estates (1744–1766)

	1744	1745	1746	1747	1761	1763	1765	1766	Total
Baulks and beams	—	1,436	564	—	729	668	4,956	957	9,310
Logs	116	—	—	117	247	—	—	—	480
Masts	—		24	161	—	14	295	3	497
Spikes	—	119	206	—	—	79	669	—	1,073
Bowsprits	32	52	85	33	—	686	90	—	978
Staves	—	2,891	99	—	—	—	4,199	—	7,189
Trees	3,180	—	180	—	—	—	139	—	3,499
Total	3,328	4,498	1,158	311	976	1,447	10,348	960	23,026

SOURCE: AGAD, WAR, KOLEKCJA XX-HANDEL RZECZNY EXP. 32, 80, 85, 112 FOR YEARS 1744–1766

logs, 1,847 masts and spikes, 280 planks, 148 timbers, 737 "Dutch trees", and 309 undefined pieces (see Table 34).[128]

The data for both periods (1744–1766 and 1767–1793) show that in the second half of the 18th century, there was a change in production patterns when beams and baulks became the main product to come out of the Radziwiłł forests, which in the second of these periods accounted for as much as 89 per cent of the entire production. In contrast, in the first period, spanning the 1740s and 1760s, production was much more diversified, and beams and baulks accounted for only 41 per cent of the harvested wood. This reflects the rapid economic development undergone by the Polish-Lithuanian Commonwealth in the second half of the 18th century, when there was a great demand for wood from the industrial sector,[129] and also for shipbuilding, with the large orders placed by Britain, France, and Spain in the 1760s, 1770s, and 1780s.[130]

---

128 It should be remembered that the data for this period is partial.

129 Marian Drozdowski, *Podstawy finansowe działalności państwowej w Polsce 1764–1793* (Warsaw: PWN, 1975); Maria Kwapień, Józef Maroszek, and Andrzej Wyrobisz (eds.), *Studia nad produkcją rzemieślniczą w Polsce (XIV–XVIII w.)* (Wrocław: Zakład Narodowy im. Ossolińskich, 1976); Zenon Guldon and Lech Stępkowski (eds.), *Z dziejów handlu Rzeczypospolitej w XVI–XVIII wieku: studia i materiały* (Kielce: Wyższa Szkoła Pedagogiczna im. Jana Kochanowskiego, 1980); Antoni Mączak (ed.), *Encyklopedia historii gospodarczej Polski do 1945 roku* vol. 1 (Warsaw: Wiedza Powszechna, 1981).

130 Paul W. Bamford, *Forest and French Sea Power: 1660–1789* (Toronto: University of Toronto Press, 1956); Ernest Harold Jenkins, *A History of the French Navy, from Its Beginnings to the*

TABLE 34   Wood types (in pieces) produced in the Radziwiłł estates (1767–1793)

	1767	1768	1769	1770	1771	1772	1774	1776	1777	1778	1779	1780	1788	1789	1790	1793
Baulks-beams	2,195	561	180	1,192	1,071	348	741	1,274	4,054	5,742	1,595	727	4,854	1,129	2,394	2,679
Logs	—	—	—	—	—	—	—	41	—	59	—	—	—	—	—	—
Masts and spikes	1,846	—	—	—	—	—	—	—	—	1	—	—	—	—	—	—
Planks	—	—	—	—	—	—	—	280	—	—	—	—	—	—	—	—
Timbers	—	—	—	—	—	—	—	148	—	—	—	—	—	—	—	—
"Dutch trees"	—	—	737	—	—	—	—	—	—	—	—	—	—	—	—	—
Undefined	—	—	309	—	—	—	—	—	—	—	—	—	445	—	—	—
Total	4,041	561	1,226	1,192	1,071	348	741	1,743	4,054	5,802	1,595	727	5,299	1,129	2,394	2,679

SOURCE: AGAD, WAR, KOLEKCJA XX-HANDEL RZECZNY EXP. 10, 31, 32, 77, 112 FOR THE PERIOD 1767–1793

TABLE 35   Number of beams and baulks (in pieces) obtained from the Radziwiłł forests divided by
length (1767–1780)

	1767	1768	1769	1770	1771	1772	1774	1776	1777	1778	1779	1780	Total
3 fathoms	—	—	—	—	—	—	—	16	38	79	4	7	144
4 fathoms	114	—	—	—	—	—	—	163	284	342	49	44	996
5 fathoms	528	—	—	—	—	4	—	574	850	1,009	280	133	3,378
6 fathoms	708	272	60	336	271	147	175	236	1,406	2,402	753	249	7,015
7 fathoms	456	138	60	415	495	52	335	169	764	1,190	353	130	4,557
8 fathoms	243	151	60	390	505	145	231	29	610	519	116	78	3,077
9 fathoms	146	—	—	—	—	—	—	1	87	114	35	84	467
10 fathoms	—	—	—	—	—	—	—	—	7	8	4	2	21
11 fathoms	—	—	—	—	—	—	—	—	—	1	1	2	2
Total	2,195	561	180	1,141	1,271	348	741	1,188	4,046	5,664	1,595	727	19,657

SOURCE: AGAD, WAR, KOLEKCJA XX-HANDEL RZECZNY EXP. 10, 31, 32, 77, 112 FOR THE PERIOD
1767–1780

Looking at beams and baulks produced in 1767–1793 (see Tables 35 and 36),
it is of note that most beams were cut to lengths of 5, 6, 7, and 8 fathoms, the
most suitable sizes for construction and shipbuilding. The largest number of
beams-baulks corresponds to the 6-fathom length, with 11,219 pieces, a large
proportion of which went to the ports of Riga, Königsberg, and Memel.[131] In
Gdańsk/Danzig,[132] in contrast, the most popular beams and baulks were those
cut to a length of 4 and 5 fathoms and, to a lesser extent, 6 fathoms, which

---

Present Day (1st UK edn, London: Macdonald & Jane's, 1973); Jan Glete, Navies and Nations:
Warships, Navies and State Building in Europe and America, 1500–1860, vol. II (Stockholm:
Almqvist & Wiksell International, 1993); Richard Harding, Seapower and Naval Warfare,
1650–1830 (London: Routledge, 1999); Olaf Uwe Janzen, Merchant Organization and Mar-
itime Trade in the North Atlantic: 1660–1815 (Liverpool: Liverpool University Press, 1998);
Nicholas A.M. Rodger, The Command of the Ocean: A Naval History of Britain, 1649–1815,
vol. II (New York: Norton, 2004); Torres Sánchez, Military Entrepreneurs and the Spanish
Contractor State in the Eighteenth Century.

131   Sven-Erik Åström, From Tar to timber: Studies in Northeast European Forest Exploitation and Foreign Trade 1600–1860 (Helsinki: Societas Scientiarum Fennica, 1988); Pierrick Pourchasse, "The Control of Maritime Traffic and Exported Products in the Baltic
Area in Early Modern Times: Eighteenth-Century Riga", Revue historique 686, no. 2 (2018),
377–398.

132   Czesław Biernat, Statystyka obrotu towarowego Gdańska w latach 1651–1815 (Warsaw: PWN,
1962), 224–225.

TABLE 36   Number of beams and baulks (in pieces) obtained from the Radziwiłł forests
            divided by length (1788–1793)

	1788	1789	1790	1791–1792	1793	Total
3 fathoms	24	3	16	No data	12	55
4 fathoms	680	116	283	No data	205	1,284
5 fathoms	1,787	378	1,003	No data	858	4,026
6 fathoms	1,758	532	835	No data	1,079	4,204
7 fathoms	494	82	219	No data	393	1,188
8 fathoms	111	17	33	No data	115	276
9 fathoms	—	1	5	No data	20	26
10 fathoms	—	—	—	No data	—	—
11 fathoms	—	—	—	No data	—	—
Total	4,854	1,129	2,394	No data	2,682	11,059

SOURCE: AGAD, WAR, KOLEKCJA XX-HANDEL RZECZNY EXP. 10, 32, PERIOD 1788–1793

proves that the wood harvested in the Radziwiłł forests was better than that
harvested along the Vistula.

11   Examples of Radziwiłł Timber Trade in Southern Baltic Ports

Riga was the main gateway for the timber purchased by the navies of Great
Britain, France, the Netherlands, Sweden, Denmark, and Spain in the southeast
Baltic.[133] By the end of the 17th century, Riga had become an important centre
for the trade in forest products, and the timber sourced from its hinterland was
approximately 25 to 30 per cent cheaper than in the main western harbours.
This was because the forests near the Daugava River and in the Grand Duchy of
Lithuania began to be exploited on a large scale in the second half of the 17th
century, unlike the forests in the Polish Crown, whose mass exploitation began
much earlier, between the 15th and the late 16th centuries. In addition, these

---

133   Pourchasse, "The Control of Maritime Traffic and Exported Products in the Baltic Area in
       Early Modern Times: Eighteenth-Century Riga", 378–381.

wood sources were located far from the main rafting rivers, which substantially increased the final price.[134]

To obtain timber, merchants from Riga sent their representatives deep inland, especially into the Daugava River basin, where these agents secured logging permits from local officials. The practice began in the 17th century, but it gained traction especially in the 18th century, when European powers began to show enormous interest in Riga's wood. This can be illustrated by two contracts: one was signed in 1728 by an Orszański huntsman, Michał Żuk, and a landowner called Wiśniewski, who owned vast forests; the contract was for the preparation of masts and bowsprits for the Riga-based merchant Barklaj de Tolli; the other was the contract signed by Rafał Ślizień and two Jewish merchants from Połock in 1781, to source wood, mainly timber and mast logs, during a six-year period.[135]

Concerning the transport of wood from the Radziwiłł forests, the main outlets, apart from Riga, were the ports of Memel and Königsberg, and to a lesser extent Kaunas and Świeżno. Interestingly, merchant records for the years 1763–1793 show that the Radziwiłł family kept the largest network of commercial associates in Königsberg, where their most famous trading partner was the Saturgus family, descendants of the famous entrepreneur Fridrich Saturgus (1697–1754), who began working with the Radziwiłłs as early as 1741. In the years 1763–1781, the Saturgus family worked closely with the Radziwiłł, selling wood, potash, and other products for them, and holding a virtual monopoly over the Radziwiłłs' goods in Königsberg. Other important figures for trade in this harbour, who also worked with the Radziwiłłs, were Józef Jabłoński, Michał Kloppmann, and Kondratowicz, as well as the Königsberg rafting commissioner between 1763 and 1794.[136]

Other Königsberg merchants employed by the Radziwiłłs in the second half of the 18th century were Joachim Mojzesz Fridlander, Salomon Hirsz, Aleksander Boris, Hirsz Jankielowicz, Toussaint Laval, Timotheus Anderson, Scheress, Bekensztein, Schindelmeisser, and others (see Table 37). Most of these traders were of Jewish origin and often had their own commercial network in the Radziwiłł estates. This is shown by documents that prove that in 1763, Friedrich II Saturgus had a network in Karol Radziwiłł's (1734–1790) estates; for

---

134   Stanisław Gierszewski, *Wisła w dziejach Polski* (Gdańsk: Wydawnictwo Morskie, 1982), 23–24.
135   Hedemann, *Dawne puszcze i wody*, 61–63.
136   AGAD, WAR, Kolekcja XX-Handel Rzeczny, exp. 10 for years 1763–1793.

TABLE 37   List of merchants in the Radziwiłł family commercial network (1741–1794)

Period/year	Name	Place
1741–1781	Saturgus family	Königsberg
1777–1779	Bekensztein	Königsberg
1778	Joachim Mojzesz Fridlander	Königsberg
1779–1785	Scheress	Königsberg
1781	Ludwik Grave	Riga
1782–1787	Aleksander Boris	Königsberg
1784	Hirsz Jankielowicz	No data
1784	Timotheus Anderson	No data
1785	Schindelmeisser	Königsberg
1787	Salomon Hirsz	No data
1789	Ogilvie	Memel
1789	Waterson	Memel
1790–1794	Rupel	Memel
1793	Robinson	Memel
1793–1794	Durno	Memel
1793	Toussaint Laval	Königsberg

SOURCE: AGAD, WAR, KOLEKCJA XX-HANDEL RZECZNY EXP. 10, 13, 15, 77 FOR THE PERIOD
1741–1794

instance, the merchant Abel and Szeresz served his interests in Kiejdany, and
Fernet and the pharmacist Szmyt in Nieśwież.[137]

These examples clearly show that the Radziwiłł family had much stronger
ties with smaller Baltic ports such as Memel and Königsberg than with the
large harbour of Riga. This was most likely due to distance and greater competition, since, as noted, Riga merchants sought trade agreements by sending
their intermediaries to the hinterland, rather than dealing with the Radziwiłł family. These data are also confirmed by records of shipped timber for
the years 1767–1788 (see Tables 38–40). The available registers for 1767–1788
suggest that the port of Königsberg was the one to which the largest number
of timber transports were sent (six transports), followed by Riga (three) and
Memel (three). Wood was also sent to other places, such as Kaunas, Swierzno,
and Mikołajewszczyzna, and the forest in which the most wood was harvested

---

137   Ibid., exp. 13 for years 1741–1763.

TABLE 38 Examples of amount of wood hauled in the Radziwiłł forests (1767–1776)

	1767	1768	1769	1770	1771	1772	1774	1776
Production centres	Vyžuonos, Dubingiai	Žahalska	Žahalska	Žahalska	Žahalska	Žahalska	Žahalska	Žahalska, Orciuchy
Destination	Kaunas	Königsberg	Swierzno, Königsberg	Swierzno	Kaunas	Mikołajewszczyzna	Königsberg	Riga
Land transport Amount of wood	587	111	180	262	300	348	No data	798
Rafting Amount of wood	131	306	357	No data	No data	No data	205	1,491

SOURCE: AGAD, WAR, KOLEKCJA XX-HANDEL RZECZNY EXP. 31, 32, 77, 112 FOR YEARS 1767–1776

TABLE 39   Examples of the amount of wood hauled in the Radziwiłł forests (1777–1788)

	1777	1778	1779	1780	1788
Production centres	Żahalska,	Żahalska, Dubingiai Königsberg,	Żahalska, Khotova	Żahalska	*ordynacja* of Nieśwież
Destination	Riga	Riga	Königsberg	Königsberg	Memel
Land transport Amount of wood	No data	No data	No data	600	4,624
Rafting Amount of wood	2,849	2,555	474	No data	No data

SOURCE: AGAD, WAR, KOLEKCJA XX-HANDEL RZECZNY EXP. 32, 77 FOR YEARS 1767–1776

TABLE 40   Examples of profits for the Radziwiłł family from the wood trade (1776–1778)

	1776		1777		1778	
Sale place	Memel		Riga		Riga, Königsberg, Memel	
Production place	Jašiūnai, Naliboki, Nieśwież		Naliboki, Khotova, Orciuchy, Żahalska		Naliboki, Khotova, Orciuchy, Żahalska, Dubingiai	
Type	**Amount**	**Price**	**Amount**	**Price**	**Amount**	**Price**
Baulks-Beams	2,776	49,271 zł[a]	3,935	73,636 zł 15 gr[b]	5,983	80,501 zł 18 gr
Spikes	60	6,000 zł	95	6,270	1	108 zł
Bowsprits	242	968 zł	101	404	—	—
Logs	3	24 zł	25	25	59	360 zł
Undefined	No data	17,176 zł 24 gr	No data	28,046 zł 6 gr	—	—
Total	No data	56,263 zł 24 gr	No data	108,556 zł 21 gr	6,043	80,969 zł 15 gr

a   zł = złoty
b   gr = grosze

SOURCE: AGAD, WAR, KOLEKCJA XX-HANDEL RZECZNY EXP. 32, 77, 112 FOR YEARS 1776–1778

was Żahalska. Importantly, some of the batches were brought overland and some down the Pregoła, Neman, and Daugava rivers.

Finally, looking at the network of trade links and locations whence wood from forests under the jurisdiction of the Radziwiłłs were shipped, it can be argued that most sales of forest products, but also cloth, wine, and weapons were smaller Baltic ports that specialised in forest goods, like Memel and Königsberg. It seems that Riga was only rarely used as an alternative to these, and not as the main gateway for the Radziwiłł estate products, which is also confirmed by the accounts for the masts ordered in the second half of the 1760s by the Spanish Navy (see Chapter 1), in which the Radziwiłłs were only one among the many contractors supplying the masts.

To close this second chapter, Map 8 illustrates timber extraction areas and its transport to the main south Baltic ports, such as Szczecin/Stettin, Gdańsk/Danzig, Königsberg, Memel, and Riga in the second half of the 18th century. It is clear that in the Polish-Lithuanian Commonwealth, timber was harvested

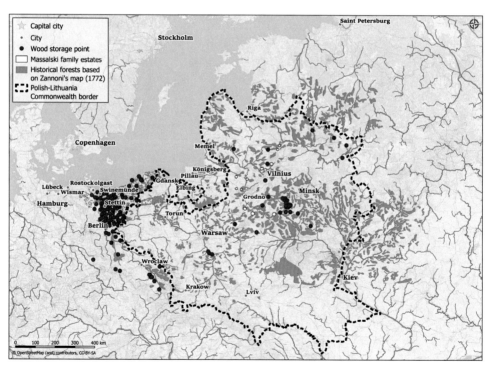

MAP 8   Places of felling and storing wood, which was later sent by the rivers to the ports of Szczecin, Gdańsk, Königsberg, Memel, and Riga

SOURCE: LOOK4GIS-LUKASZ BRYLAK BASED ON QGIS SOFTWARE

in inland forests and then shipped to the coast down the main rivers (Vistula, Bug, Niemen, Priegola, and Daugava). The situation was different in Szczecin/ Stettin, which mainly used the forest resources of Pomerania and the Szczecin Bay, and only occasionally timber rafted down the Oder from the regions in the interior.

PART 2

..

CHAPTER 3

The Survey of New Spain's Woodlands

In contrast to Chapters 1 and 2, this one addresses the exploration of forest resources in the Viceroyalty of New Spain—an overseas territory directly under the authority of the Spanish Crown—carried out by the viceregal administration and military officers. The issue reveals Spain's interest in exploiting forest resources in the different regions of the Gulf of Mexico and the Caribbean Sea within the framework of the naval development policies unleashed in the second half of the 18th century.

These efforts resulted in a better understanding of the forest resources of the provinces of Veracruz, Oaxaca, Laguna de Términos, and Tabasco, as well as Louisiana and Cuba, and led to logging operations in places like Chimalapas, where pines for masts were cut under the patronage of the Crown, as well as the granting of the first *asientos* for the harvesting of cedar and mahogany near the Coatzacoalcos River and the settlement of Tlacotalpan. Like Chapter 2, historical documents are interrogated for the identification of logging sites used to meet the needs of the Spanish Navy.

1 The Exploitation of Timber for the Shipbuilding Industry in the Viceroyalty of New Spain in the 18th Century

During the reign of Philip V, several figures of the Spanish Enlightenment, such as Francisco de Seijas y Lobera,[1] Jerónimo de Uztáriz,[2] and José del Campillo y Cossío[3] described different aspects of the administration of the Spanish Empire and made numerous proposals for fiscal, commercial, administrative,

---

1 Francisco de Seijas y Lobera's *Gobierno militar y político del reino imperial de la Nueva España*, is the seventh of 14 volumes written under the overall title *Memoria sobre el gobierno de las Indias españolas realizada por don Francisco de Seijas y Lobera para servir a la verdadera unión de las dos coronas de España y Francia* (published between 1702 and 1704).
2 Jerónimo de Uztáriz, *Theorica y practica de comercio, y de marina: en diferentes discursos y calificados exemplares, que con específicas providencias, se procura adaptar á la Monarquia española*, book dedicated to Philip V (published in 1724).
3 José del Campillo y Cossío, *Nuevo sistema de gobierno económico para la América: con los males y daños que le causa el que hoy tiene, de los que participa copiosamente España; y remedios universales para que la primera tenga considerables ventajas, y la segunda mayores intereses* (written in 1743 and published in 1789).

© KONINKLIJKE BRILL BV, LEIDEN, 2024 | DOI:10.1163/9789004689640_005

political, and social reform. Concerning trade reform, all of them gave priority to reinforcing the *Real Armada*, whose task was to protect and uphold commercial exchange with the American colonies.[4] Their works emphasised the importance of American forest resources, which were both abundant and rich in excellent timber, especially cedar and mahogany, which could be used by the Spanish naval industry to build new ships-of-the-line, frigates, and smaller vessels, and thus play their part in the Spanish naval revival. A good example of this is the following passage of Seijas y Lobera's *Gobierno militar y político del reino imperial de la Nueva España* (1702):[5]

> if to build ships and powerful fleets in both seas [the South and Northern seas] timber is needed, there is no better country in the world for it than New Spain, because, like in the other countries in the [West] Indies, cedar and mahogany are found in both coastlines to make the ship hulls, and for the masts there are guayacanes, pine trees and other trees, like cugues and marías.[6]

Although Seijas y Lobera's proposal was rejected by Philip V, the idea of using American forest resources stayed alive and grew over time, as reflected by the important economist and influential merchant Jerónimo de Uztáriz, whose *Theorica y practica de comercio, y de marina* argued that:

> a great asset for Your Majesty is the abundance and quality of timber, pitch, and tar in American islands and continent [...] and it is a great advantage because if [ships] built in Europe last twelve to fifteen years, [American ones] go on for 30, being made with the hardiest cedar and other superior woods; this also means that they need less maintenance

---

4 See Geoffrey J. Walker, *Spanish Politics and Imperial Trade, 1700–1789* (London: Macmillan, 1979); David B. Bradind, *Bourbon Spain and Its American Empire* (Cambridge: Cambridge University Press, 1981); Henry Kamen, *Spain's Road to Empire: The Making of a World Power, 1492–1763* (London: Allen Lane, 2002); María Baudot Monroy, *La defensa del Imperio. Julián de Arriaga en la Armada (1700–1754)* (Madrid: Ministerio de Defensa-Universidad de Murcia, 2013); Adrian J. Pearce, *The Origins of Bourbon Reform in Spanish America 1700–1763* (New York: Palgrave Macmillan, 2014).

5 He was in America between 1692 and 1701.

6 si para fabricar en ambos mares [del Sur y Norte] poderosas armadas se requiere en primer lugar las maderas, no hay países en todo el mundo más aptos para ellas que los de la Nueva España, porque como en los demás de aquellas Indias, se hallan en los puertos y costas de ambos mares las famosas y permanentes maderas de cedro y las de las caobas para hacer los cascos de los bajeles, y para arbolarlos los guayacanes, los pinos y otros árboles como son los cugues y marías. (Seijas y Lobera, *Gobierno militar y político*, 293).

and repairs. That is apart from the fact that in combat cedar absorbs bullets without splintering, which in European ships kills and wounds many.[7]

Unfortunately, this thinker's interest in intensifying the exploitation of American woods was, again, ignored by state officials. However, the idea did not go away, and came up time and time again in other reformist works, such as *Nuevo sistema de gobierno económico para la América* (1743), written by José del Campillo y Cossío, ex-governor of Cuba: "cedar, mahogany, and other beautiful woods, but also ship masts, boards, pitch, tar, and other materials that we currently buy from the Baltic, are available at the Indies [...] and we could take them there and sell them cheap".[8]

These efforts to promote the use of American woods in royal ships could not be brought to fruition because of the opposition of the traditionalists, led by Antonio Gaztañeta, who saw these attempts to encourage naval construction in the Indies with the support of José Patiño as a threat to the Spanish establishment.[9] In addition, previous experiences in New Spain, like the creation of a major shipyard in Coatzacoalcos and another one for smaller ships in Campeche had met with little success.[10] The first attempt to create a royal shipyard, in Coatzacoalcos, aimed to make use of the region's wood and also the city's easy access to the mountainous regions of Oaxaca and Veracruz, where pine trees for masts were to be found in abundance. The earliest proposal to build a shipyard there was floated in 1701 by the admiral of the Armada

---

7   son grandes las ventajas que en las islas y tierra firme de la América tiene su majestad de muchas y exquisitas maderas y abundancia de brea y alquitrán para la construcción de bajeles [...] con el considerable beneficio de que si los [buques] fabricados en Europa duran de 12 a 15 años, [estos de América] se conservan más de 30, ya que se hacen allá con el cedro, roble más duro y otras maderas de superior firmeza y resistencia; lo que es causa también de que necesitan de menos carenas y de otros reparos. Fuera de que en un combate tiene también el cedro la ventaja de que embebe en sí las balas, sin que se experimenten los efectos de los astillazos, que en los navíos fabricados en Europa, y que suelen maltratar y aun matar mucha gente. (Uztáriz, *Theorica y practica de comercio*, 216).

8   cedro, caoba y otras maderas hermosas, pero también mástiles para navíos, tablazón, brea, pez y otros géneros gruesos que ahora nos vienen del Báltico, los tendremos de nuestras Indias [...] y los podremos llevar allá y venderlos baratos. (Campillo y Cossío, *Nuevo sistema de gobierno económico*, 158).

9   Valdez Bubnov, *Poder naval y modernización del Estado*, 204.

10   Between 1702 and 1721, the shipyard in Campeche built a minimum of two ships for the Real Armada. The first was the 50-gun *Nuestra Señora de Guadalupe*, launched in 1702, and the second was the *La Potencia del Santo Cristo*, alias "El Blandón", launched in 1720. Amparo Moreno Gullón, "La Matrícula de Mar de Campeche (1777–1811)", *Espacio, Tiempo y Forma, serie IV, Historia Moderna* 17 (2004), 275.

de Barlovento, Don Guillermo Morfi, but nothing was done at the time. After-wards, the idea was revisited by viceroys Marquis of Valero (1716–1722) and Marquis of Casa Fuerte (1722–1734), but it was only the latter that could see the project moving ahead, when José Patiño authorised the construction of two ships-of-the-line in Coatzacoalcos in 1726. However, the project was beset with problems from the start, including disagreements with the Creole elite and with naval officials in Havana, who, in essence, refused to send ship carpenters to Mexico. For these reasons, the project was downsized to the construction of a single ship, the 60-gun *San José*, alias *"Nueva España"*, with a 66-cubit keel, built over three years (1732–1735) at enormous cost for the *Real Erario* (431,000 pesos de a ocho). Finally, the government, in view of the administrative and logistic problems encountered and the high cost, decided to cancel the project and continue ship construction in Havana.[11]

It is worth pointing out that, while these projects were in the planning stage, two surveys were undertaken to identify woodland that could be used by the naval industry. One was carried out in 1726 by the navy's general lieutenant, Antonio Serrano, in the forests of the isthmus of Tehuantepec, where he spot-ted good pine trees and other useful species. The second survey, by squadron leader Rodrigo de Torres, examined woodland in the region of Coatzacoalcos in 1733, following the complaints filed by Juan Pintado and Pedro de Torres at the beginning of the construction of the *Nueva España* in the shipyard situated on the river mouth.[12]

Following these failed attempts, the next naval construction programmes for the *Marina Real* could not be brought forward until the 1760s. The change of policy was triggered by Spanish defeat in the Seven Years' War (1756–1763), in which Spain had been involved since January 1762, after the signature of the III Family Pact by French and Spanish Bourbons on 15 August 1761. The occupa-tion of Havana and Manila and several territorial losses shocked Charles III's government, who immediately ordered reforms to reinforce the navy and the army in both Spain and the colonies. This period, which began in 1765 and lasted until the king's death in 1788, was crucial for military reform in Spain, especially concerning the *Armada Real*. The reorganisation of the service in the 1780s turned Spain into the second naval power in Europe.

---

11   Germán Andrade Muñoz, *Un mar de intereses* (México: Instituto Mora, 2006), 85–87; Valdez-Bubnov, *Poder naval y modernización del Estado*, 240–241.

12   Archivo General de la Nación de México, Reales Cedulas Originales, vol. 53, fs. 122–123; Antonio Béthencourt Massieu, "Arboladuras de Santa María de Chimalapa Tehuantepec en las construcciones navales indianas 1730–1750", *Revista de Indias* 20 (1960), 70–72; John T. Wing, *Roots of Empire: Forests and State Power in Early Modern Spain, c.1500–1750* (Leiden: Brill, 2015), 195.

One major change was that the Spanish ceased building ships "English style", as was still done while the Marquis of Ensenada was in office, and began constructing them "French style". In 1765, the French shipbuilder Francisco Gautier was put in charge of ship design and the supervision of shipyard work. In addition to this, in 1772 Gautier wrote an important report entitled *Observaciones sobre el estado de los montes de España, nota del consumo de la madera de construcción, que, en cada año, se considera necesaria en los departamentos de Ferrol, Cartagena y Cádiz; y proyecto para aprovisionar estos arsenales de maderas de América*, in which he described the state of Spanish woodlands and suggested the use of American woods in Iberian shipyards: "Cádiz uses American wood which is brought at great cost but with little profit, not because it lacks in quality, but because of the carelessness with which it is dispatched [to the metropolis], where it arrives badly cut, badly arranged, and poorly sorted".[13] In this passage, Gautier was saying more than it seems, because what he was really suggesting was to shift the weight of wood supply from Spanish shipyards to the American forests; as he pointed out, there was an exchange of wood between Havana and other locations, but this fell short of requirements and was limited to very specific ship parts. This is also reflected in a report issued by Ciprián Autrán, shipbuilder in La Carraca-Cádiz, in 1749, *Estado de las piezas de sabicú y caoba que se necesitan para la construcción de un navío de 70 cañones y otro de 80*. The engineer pointed out that the sabicú, being harder than oak, was ideal for keels, keelsons, sternposts, stems, mast bases, brackets, yokes, and tillers. He also mentioned mahogany, which was in his opinion ideal for rudders, beams, drums, and bitts.[14]

Gautier's idea to source all timber for naval construction in American forests aimed to preserve Spanish woodland resources, so that they could be used during emergencies or wartime, when navigation routes and the supply of strategic materials were interdicted. Eventually, in March 1767, the Secretary of the Navy, Julián de Arriaga, heeded Gautier's advice and ordered the examination of pine and cedar sheathing to confirm which one was more resistant to shipworm. The test was undertaken with two royal frigates, *La Flecha* and

---

13   "Cádiz utiliza la madera de América que viene a mucha costa y de la cual se saca muy poco provecho, no por su calidad, que es de preferir a cuanto roble hay en España, sino por el ningún cuidado de los que la envían a [la metrópoli], donde llega muy mal delineada, mal configurada y nunca surtida". AGS, SMA, Arsenales, leg. 349, Observaciones sobre el estado de los montes de España, nota del consumo de la madera de construcción, que, en cada año, se considera necessaria en los departamentos de El Ferrol, Cartagena y Cadiz; y proyecto para aprovisionar estos arsenales de maderas de America.

14   AGS, SMA, Arsenales, leg. 317, Estado de las piezas de madera de sabicú y caoba que se necesitan para la construcción de un navío de 70 cañones y otro de 80.

La Perla. The officials Pedro de Acosta, Juan de Mora, Joseph Chenard, Vicente Morand, Salvador del Castillo, and Pedro Prieto reported that the pine sheathing, in the former, lasted 15 months, and the one made of cedar, in the latter, 21: "cedar is more durable, as ships that were sheathed [with it] here went to Spain and back with the same sheathing".[15] The officials also reported that the main problem with pine was the woody structure, because shipworm attacked the sapwood, thinning out the boards. Cedar, they pointed out, was also attacked by shipworm, but the internal structure was not affected, and the sheathing did not become thinner.[16]

This study, which was undertaken at the *maestranza* of Havana, was well received by the Secretary of the Navy and was a turning point in the use of American forest resources for naval construction. In the 1760s, Cuba was still regarded as the main warehouse of tropical woods, as shown by a report issued by the Count of Macuriges about tree felling on the island between 1766 and 1769,[17] but by the end of the decade, the first initiatives to use resources from New Spain began to come to fruition.

2   Pine Felling in the Hills of Chimalapas

The woodlands of Chimalapas, in the isthmus of Tehuantepec, were an important source of wood—cedar, pine, and oak—for shipbuilding from the early years of the colony. The area was regarded as especially suitable for this purpose owing to its strategic location in the corridor that links the Gulf of Mexico and the Pacific Ocean. The Coatzacoalcos River, which runs down from the hills to the Gulf, was used to haul goods and, especially, timber, and the Zoque

---

15   "El cedro es de más duración, pues hemos visto, que navíos que se han forrado aquí y han ido a España, han vuelto en los mismos foros".

16   AGS, SMA, Arsenales, leg. 317, Se contexta el Real Orden expedido, para que el forro exterior de los navíos sea de tablas de pino; y se expone, con dictamen de la Maestranza ser esta calidad de madera de muy corta duración según experiencias.

17   Ibid., leg. 343. The intendent of the department in Havana reported that contractors had delivered 33,247 pieces for the construction of approximately eight 60- to 70-gun ships; 2,142 for frigates; and 697 for *xebecs* for the coastguards; while another 3,000 pieces were still to be brought to the port. In 1769, Francisco Mendoza, navy officer, and Joseph Chenar, building assistant, arrived in Havana to inspect the island's woodland and supervise woodcutting operations. They were asked to supply 24,000 pieces of different sorts, for 60- to 120-gun ships. Rafal Reichert, "Recursos forestales, proyectos de extracción y asientos de maderas en la Nueva España durante el siglo XVIII", *Obradoiro de Historia Moderna* 28 (2019), 63.

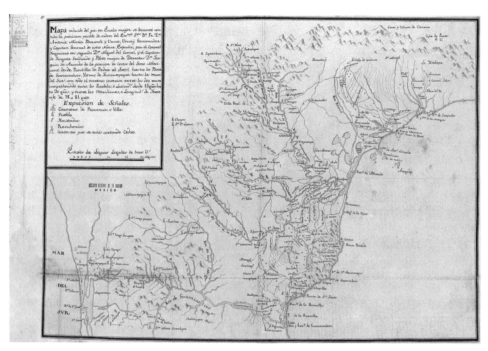

FIGURE 7   Map from 1793 showing the layout of mountains and rivers in the provinces of Veracruz and
Oaxaca. With marked cedars felling sites. Made by Miguel del Corral & Joaquín de Aranda
SOURCE: ARCHIVO GENERAL DE LA NACIÓN DE MÉXICO, MAPAS, PLANOS E
ILUSTRACIONES (280), MAPILU/210100/396

villages of Santa María and San Miguel, both in Chimalapas, whose inhabitants
regarded themselves as the ancestral owners of this vast woodland area, played
a prominent role in its exploitation.[18] From an administrative perspective, the
region depended on the Alcaldía Mayor of Tehuantepec, but Nexapa (modern Nejapa de Madero, Oaxaca) also had an important part to play, especially
by supplying indigenous labour and paying the salary of woodcutters and day
labourers, for instance during the felling of trees intended for the masting of
royal ships in 1768.[19]

---

18   Laura Machuca, "Proyectos oficiales y modos locales de utilización del Istmo de Tehu-
      antepec en la época colonial: historias de desencuentros" in Emilia Velázquez and Éric
      Léonard and Odile Hoffmann and M.-F. Prévôt-Schapira (eds.), *El istmo mexicano: una
      región inasequible. Estado, poderes locales y dinámicas espaciales (siglos XVI–XXI)* (Mar-
      seille: IRD Éditions, 2009), 66.
19   AGNM, Industria y Comercio, vol. 10, exp. 1, fs. 46–47, "Corte real de arboladuras en Chi-
      malapa".

Tree felling in the woodlands of Chimalapas intensified in the 18th century, when the Bourbon reforms and the creation of the Secretary of the Navy and the Indies in 1714 led to plans to build a shipyard in the bar of Coatzacoalcos and promote shipbuilding to south of Veracruz. The project failed, but the excellent quality of the region's woods, especially pine for masts, was not lost on Spanish shipbuilders. In 1735–1736, 1738, 1741–1743, and 1747–1748, limited tree-falling campaigns were carried out to supply the shipyard in Havana, through Veracruz, with timber for masting. Interestingly, these campaigns were preceded by the press-ganging of woodmen in Tlacotalpan and day labourers in Tehuantepec, Nexapa, and Guamelula (modern San Pedro Huamelula, Oaxaca), because nobody wanted to take such demanding and badly paid jobs (in 1735–1736, for instance, the daily wage was only 2 reales).[20] It must be pointed out that only the first of these campaigns yielded satisfactory results, as the others were beset with administrative incompetence and logistic issues when it came to floating the timber down to the mouth of the Coatzacoalcos. Some of the wood became unusable as a result, and tree-felling campaigns in the area did not resume until January 1765, when the Navy Intendent of Havana requested new timber for masting. This was the last time the region supplied pinewood to this department.

3   Pine Felling in Chimalapas under Royal Supervision (1765–1771)

The idea to resume tree felling in Chimalapas came up in the discussions held by Lorenzo Montalvo Avellaneda and Juan Gerbaut, navy intendents in Havana and Cádiz, and the Viceroy of New Spain, Joaquín Juan de Montserrat y Cruillas, in which the point was made that "the Coatzacoalcos pines[21] cost just over half as those that come from the north, and although they are of a lesser quality, being of heavy wood, they are fairly good for masting".[22] For good measure, the intendent in Cuba, Lorenzo Montalvo Avellaneda, emphasised that it was much cheaper to source the timber there than having the masting sent from

---

20   Béthencourt Massieu, "Arboladuras de Santa María de Chimalapa Tehuantepec", 65–101.
21   "Coatzacoalcos pines" is a recurrent expression in the record, which alludes to the place where they were loaded onto the waiting ships; i.e. the bar of Coatzacoalcos, in the Gulf of Mexico. These pines, however, were in all probability sourced in the woodlands of Chimalapas.
22   "pinos de Coatzacoalcos salían a poco más de la mitad del costo que le tienen al Rey, los pinos del Norte, que aún de inferior calidad que estos, por ser a mayor peso su madera, son de bastante bondad para arboladuras de navíos". AGS, SMA, Arsenales, leg. 339, f. 28, Pinos de arboladura de Goatzacoalcos.

the Spanish warehouses in Cádiz and El Ferrol. Eventually, the project began moving forward in early 1765, and by 25 June a total of 223 trees had been felled, for a total cost, including haulage to the bar of Coatzacoalcos, of 27,099 pesos de a ocho reales. In his report, the viceroy mentions that 26 masts were sent to Havana aboard frigate *Águila*, three were sold to private companies in Veracruz, 13 were loaded onto royal ships, 114 were stored in the port of Veracruz, and 67 were still in the river mouth.[23] The three officials agreed to create a permanent *tinglado*[24] for masts and cedar and mahogany parts in Veracruz, to meet the needs of navy ships and also to sell to merchant ships (see Figure 8).

It is now worth lingering on some technical aspects related to the pieces of masting requested by the Spanish Royal Navy, in order to better understand the features and size of the trees sought for this purpose. No precise information exists about the length of timber cut in earlier campaigns,[25] but in a document dated 1767, naval officer Antonio de Gregorio included a note about "the provision of trees to make masts, yards, topmasts, and outriggers for 80-, 70- and

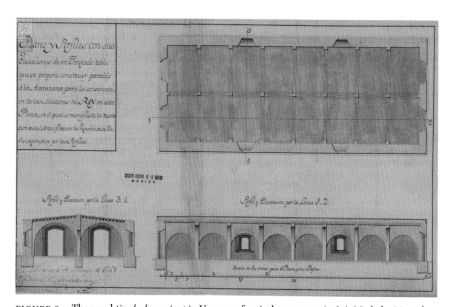

FIGURE 8   The royal *tinglado* project in Veracruz for timber storage (1764). Made by Miguel de Santiestevan

SOURCE: ARCHIVO GENERAL DE LA NACIÓN DE MÉXICO, MAPAS, PLANOS E ILUSTRACIONES (280), MAPILU/210100/417

---

23   AGS, SMA, Arsenales, leg. 339, f. 8.
24   A solid wooden or masonry structure where timbers and carved pieces were stored.
25   Andrade Muñoz, *Un mar de intereses*, 99–105.

TABLE 41   Length of timber for masting from Chimalapas, according to Antonio de Gregorio (in cubits)

Type of piece	80-gun ship	70-gun ship	60-gun ship	30-gun frigate
Mainmast "Palo mayor"	48–49	44–46	42–43	43
Foremast "Palo de trinquete"	44–46	41–43	40–41	39
Mizzen "Palo de mesana"	44	43	41	36
Bowsprit "Palo de bauprés"	38	36	34	26
Main topmast "Masteleros de gavia"	36	35	33	25
Topgallant "Mastelero de juanete"	25–26	25	20	19
Ring-tail boom "Botalones de alas"	27	26	24	19
Jibboom "Botalón de foque"	26	25	24	16

SOURCE: AGS, SMA, ARSENALES, LEG. 624, FS. 87–90

60-gun ships and also for two 30-gun frigates",[26] in relation to the felling campaigns undertaken between 1760 and 1770 (see Table 41).[27]

On the other hand, a shipment of pieces for masting intended for Havana, loaded aboard two royal packet-boats—the *San Lorenzo*, with 28 pieces, and the *San Francisco de Paula*, with 37—in June 1769, shows that only five out of 65 were over 40 cubits long (the longest measured 44); 33 of the masts, that is, over half, were between 30 and 39 cubits long; finally, 27 were between 20 and 29 cubits long.[28] Based on this, it can be deduced that most masts were midsized, and were suitable as frigate masts and as small parts for larger ships, such as topmasts, outriggers, fishes, spritsails, topgallants, and studding. This suggests that the shipyard in Havana was demanding exactly this kind of timber to conclude the construction of a frigate (*Santa Lucía*) and three ships-of-the-line (*San Francisco de Paula*, *Santísima Trinidad*, and *San José*).[29]

Returning to tree-felling operations in Chimalapas, the second campaign came to an end in late October 1766. During the rainy season, 220 pine trunks were sailed downriver to the estuary of Tacojalpa and thence to the bar of Coatzacoalcos, and 250 more were left in the hills. Some broke while they were

26   "la provisión de árboles para elaborar palos, vergas, masteleros y botalones de navíos de 80, 70 y 60 cañones y también para dos fragatas de 30 cañones".

27   AGS, SMA, Arsenales, leg. 624, fs. 85–90, Instrucción de Antonio de Gregorio para el corte de arboladuras en Chimalapa.

28   AGNM, Industria y Comercio, vol. 10, exp. 1, fs. 31–33.

29   José Manuel Serrano Álvarez, *El astillero de La Habana en el siglo XVIII: historia y construcción naval (1700–1805)* (Madrid: Ministerio de Defensa, 2018), 530–531.

being dragged and others were also left unusable for different reasons.[30] The third campaign took from 5 September 1767 to 2 April 1769. Havana officials appointed Don Juan Bautista Franco as shipbuilding supervisor and mast master, while Lieutenant Don Joseph Ximenez was commissioned to oversee tree felling and to survey the hills near Mijes, Alvarado, and Coatzacoalcos,[31] with the aim of finding additional timber with which to build two schooners; it was urgent to dispatch 67 of the 220 masts cut during the second campaign, which were still sitting in Coatzacoalcos. Overall, between 1767 and 1769 a total of 432 pines were felled for masting, and another 78, left from the 250 trees felled between 1765 and 1766, were finally fitted.[32]

Importantly, the best season for felling depended on weather conditions, tree density, type of tree required, terrain conditions, and the physical conditions of the workers. A report issued by Don Juan Antonio Ytuarte provides detailed dates for the pines felled between 20 December 1768 and 1 April 1769 (fourth felling campaign):

– up to 20 December 1768, 17 pines were made ready;
– up to 1 January 1769, 16 masts were fitted;
– up to 8 January, 32 pines were made ready;
– up to 15 January, 36 masts were fitted;
– up to 28 January, 41 pines were made ready;
– up to 5 February, 14 masts were fitted;
– up to 12 February, 41 pines were made ready;
– up to 19 February, 41 masts were fitted;
– up to 25 February, 30 pines were made ready;
– up to 5 March, 23 pines were made ready;
– up to 11 March, 12 masts and 3 bowsprits were fitted;
– up to 18 March, 40 pines were made ready;
– up to 22 March, 40 masts were fitted;
– up to 1 April, 46 masts were fitted.[33]

After the trees were felled, they were fitted, dragged to the riverside, put in the water, tied in rafts, and driven by bargemen to the bar of Coatzacoalcos. As they floated downriver, some broke against rocks or ran aground. Juan Antonio de Ytuarte gives an account of one of these incidents: "on 16 September [1769] the last of the timber was put in the water, but as we were half-course down five small masts broke; and one of the barges was wrecked in a torrent

---

30   AGNM, Industria y Comercio, vol. 10, exp. 1, f. 43.
31   AGNM, Marina, vol. 26, fs. 232–233v.
32   AGNM, Industria y Comercio, vol. 10, exp. 1, f. 19.
33   AGNM, Industria y Comercio, vol. 10, exp. 1, f. 19.

on the 31st day of last month; there were 10 men on the barge, and they were in danger because the current sent them under a raft, but they were all saved and only the barge and some tools were lost".[34] The same official also reported an accident on 22 November 1769, when a shipment was ready to go in Malpaso de Guichicovi—i.e. the rafts were already tied up—a river wave broke the ropes and some of the timber was lost in the nearby islets. This incident forced shipments to stop for a few days, and the operatives ran out of supplies. They sent a courier to communicate this unpleasant news to Don Juan Antonio de Ytuarte, who reacted by forcing Sergeant Joseph Ybañez to return to the felling area with all single men, "considering that the season was advanced for the task at hand".[35] Married men were allowed to go to Tlacotalpan and return to their families. The jam was eventually cleared, and the last masts arrived in Coatzacoalcos in February 1770. The final felling campaign that year ended in November, when 209 trees were felled and sent downriver.[36]

The total cost of the timber felled in Chimalapas was over 202,000 pesos de a ocho reales, although the most remarkable thing is that, as late as July 1773—that is, nearly two years after the felling—350 masts of different sizes and types remained in the mouth of the Coatzacoalcos, according to the evaluation undertaken by Manuel Pérez.[37] This shows that felling operations went at a brisk pace, and that the problem was at the river mouth, where the timber was often left waiting to be taken to Veracruz and Havana for a long time. These logistic difficulties were recurrent, owing to a scarcity of freight ships large enough to accommodate this masting.

4   Expenses: Wages and Oxen for the Chimalapas Enterprise

In June 1767, Don Joseph Ximenez requested more resources from the royal officials in Veracruz to continue felling in Chimalapas. The money was sent with Don Juan Antonio Ytuarte, who, on his arrival in late August, was put in

---

34   el día 16 de septiembre [1769] acabé de echar al agua toda la madera, y me hallé con ella a medio andar del río, en este tránsito se han desgraciado cinco palos de poca consideración, con una de las canoas que el 31 del pasado se hizo pedazos en un raudal, iba equipada con 10 hombres, quienes se liberaron de este lance en que se vieron con bastante peligro, por haberlos metido la corriente bajo de una partida de palos y solo se perdieron la canoa y algunas herramientas. (AGNM, Industria y Comercio, vol. 10, exp. 1, fs. 39–40)

35   "considerando lo avanzado del tiempo para los sucesivos trabajos" (Ibid., fs. 54–56).

36   Ibid., f. 68.

37   AGS, SMA, Arsenales, leg. 351, fs. 22 and 24, Relación de arboladuras en Goatzacoalcos por Manuel Pérez.

charge of the operation. The expenses incurred during the third felling campaign, which ran between 5 September 1767 and 19 October 1768, were as follows:

- 3,459 pesos, 7 tomines, and 8 granos in daily supplies for the woodmen;
- 31,042 pesos, 5 tomines and 2 granos in wages;
- 1,705 pesos and 1 tomín in shipments;
- 1,349 pesos in tackle;
- 106 pesos in medicines;
- and 27 pesos in the chapel.[38]

The total cost came to 37,690 pesos, 1 tomín, and 10 granos, most of which went in wages. It is worth mentioning that most of the workers came from the indigenous villages under the *Alcaldía Mayor* de Tehuantepec, which also paid for their salaries—on the viceroy's orders—while the rest of the costs were shouldered by the *Alcaldías* of Guamelula and Nexapa,[39] much to the annoyance of the *alcaldes mayores*, who had to use funds that had been earmarked for other purposes.

Surveying, felling, and dragging were expensive operations, apart from the fact that, during the transport of the trees to the riverside, many hands ran away, because this was the most demanding job, and nobody wanted to do it. Don Juan Antonio Ytuarte described the stratagems used to retain the men, who drew pay from the moment they left their house to when they returned: "they were paid three reales per league [of road]; three reales in working days and one in feasts and rainy days in which no work was done; as they were replaced every fifteen days, more money was spent in days in which no work was done than the opposite. This was because their villages were far away and because in this country [the woodlands of Chimalapas] it is raining all the time".[40] The supervisor also noticed that the locals did not turn up during the best time for felling, because this coincided with pressing agricultural tasks that could not be postponed. He thus asked the viceroy, Marquis of Croix, to mobilise "250 prisoners for dragging and 25 guards to keep an eye on them, as well as four carpenters, a water carrier, and an ironsmith, [so] the daily cost for food and clothing will be 1 real per man per day, and they will be ready

---

38   AGNM, Industria y Comercio, vol. 10, exp. 1, f. 12.
39   For example, in 1769 the *alcalde mayor* of Nexapa sent 9,000 pesos, and the other one from the *Alcaldía Mayor* of Tehuantepec 6,185 pesos, 3 tomines, and 6 grains.
40   se les pagaba tres reales por cada diez leguas [de camino]; tres reales el día que trabajaban y uno el que por día de fiesta o agua no trabajaban, que relevándose unas a otras de quince en quince días, más importaban los días de camino y los durante que no trabajaban que los de trabajo. Ello por la distancia de los pueblos y las continuas aguas de este país [es decir de la selva de Chimalapas]. (AGNM, Industria y Comercio, vol. 10, exp. 1, f. 17)

to work when they are told to do so; the advantage of this is such that if one masting now costs 14,000 pesos, in this way it will cost no more than 5,000".[41]

The document carried another suggestion to save the *Real Hacienda* some money in wages. Not all of the Tlacotalpan woodmen were required when the river was running low—between October and May—so Juan Antonio Ytuarte recommended that only 16, chosen among the most experienced of them (out of a total of 60), were hired during that period, to prepare and dress the timber while the others returned home. The barges could also remain in Coatzacoalcos, which would also save money. The barges and the remaining woodmen could return in the rainy season (in June) to send the timber, which would be ready to go by then, downriver. Ytuarte's request for prisoners could not be met and the following season he relied again on hired labour from the *Alcaldías Mayores* of Tehuantepec, Guamelula, and Nexapa, but in the 1768 campaign the workforce was cut down, reducing the cost for the *Real Hacienda*.[42]

The royal treasure also had to shoulder the cost of buying oxen to pull the timber. In Chimalapas, it was estimated that 48 animals, working in pairs, would be required during the felling season. The oxen were to be supplied by the *corregidor* of Jalapa, Don Bartolomé Bejarano, Marquis of Tehuantepec. In the 1768 season, not all the animals provided were found suitable, as pointed out by the timber supervisor, Juan Bautista Franco, who accused Don Juan Antonio Ytuarte of buying so-called "cabestros", which were good only to drive cattle.[43] These and other accusations—the supervisor also denounced the way tar and food supplies were handled—were not pursued by the authorities but are an illustration of malpractice and corruption in major royal enterprises, often for the benefit of local elites (in this instance in Oaxaca and Veracruz). In a letter dated 10 May 1773, the *corregidor* of Jalapa, Don Bartolomé Bejarano, presented his accounts in response to the viceroy's request, claiming that, during the felling campaigns carried out between 1768 and 1772, the administrator Juan Antonio Ytuarte had left 107 oxen in his care:

– 74 oxen, which could not be sold locally, were sent to the city of Oaxaca and were sold by the regidor Don Diego de Villa Santa for 860 pesos and 3 reales;

---

41   para esas faenas de los tiros y arrastres 250 forzados con 25 hombres de tropa para su custodia, incluyendo en este número de jurados cuatro carpinteros, un jarrero y un herrero, [así] el costo que tendrá diariamente cada uno para la comida y vestuario será de un real y se hallarán prontos a trabajar cuando se les mande, cuya ventaja es tan considerable que si ahora una Arboladura cuesta catorce mil pesos, por este medio se conseguirá con cinco mil. (Ibid.)

42   Ibid., f. 18.

43   Ibid., fs. 51–52.

- an ox found by the people in the indigenous village of Totolapa, in the juris-diction of Nexapa, was sold to Don Joseph López for 10 pesos;
- 13 oxen, which were too old and were in danger of dropping dead on the road to Oaxaca, were slaughtered and sold for 120 pesos and 2 reales, that is, 9 pesos and 2 reales each;
- four oxen died on arrival on the *hacienda* of Chicapa, of Don Bartolomé Bejarano;
- two oxen escaped from the *hacienda* to the potrero in this village, and they were impossible to catch, joining the cimarron livestock;
- seven oxen died in the potrero, two of them killed by "lions" and "tigers";[44]
- six oxen died on the road to Oaxaca.

Importantly, 312 pesos were discounted to pay for hands, auxiliary oxen, and a foreman, Anastasio Gutiérrez, who looked after the animals in the potrero nearby the *hacienda* of Chicapa and later drove them to the city of Oaxaca. The final accounts indicate that, with these sales, the *Real Hacienda* made 778 pesos and 5 reales, which means that, compared with the original expense, a cost of 1,337 pesos and 6 reales, the Royal Treasury lost 559 pesos and 1 real.[45] These accounts show that the investment in equipment, tools, and animals in military and naval projects was costly but necessary to uphold the Spanish military effort in the Gulf of Mexico and the Caribbean.

Finally, it is worth pointing out that the engineer Miguel del Corral also vis-ited Santa María de Chimalapa during his 1776 and 1777 surveys. In the *Relación* that he wrote to give an account of his visits, he claimed that:

> many pine trees were still to be found [...] but hereafter they will be harder to use because they are more out of the way. It is of well-known quality, much masting having made out of it in Havana, and there is still a good deal of it in the esteros of Postmetacan, Tacojalpa, and Tacojalpilla, those that could not be recognised because they were hidden and cov-ered in *lechuguilla*.[46]

---

44   Probably, the official is referring to Mesoamerican felines, such as the jaguar.

45   AGNM, Industria y Comercio, vol. 10, exp. 26, fs. 4–5, Arboladuras de Chimalapa. Sobre yuntas de bueyes.

46   se halló haber todavía abundancia de pinos [...]; pero los que en adelante se sacasen serán algo más costosos por hallarse ya más retirados. Su calidad es bien conocida respecto haberse conducido muchas arboladuras a La Habana, de las que todavía queda porción en los esteros de Postmetacan, Tacojalpa y Tacojalpilla, los que no se pudieron reconocer bien por hallarse muchos sumergidos y cubiertos de lechuguilla. (Alfred H. Siemens and Lutz Brinckmann "El sur de Veracruz a finales del siglo XVIII. Un análisis de la relación de Corral", Historia Mexicana 26, no. 2 (1976), 315)

This description suggests that wood was becoming scarcer in the riverside and the region gradually deforested. The woodlands of Chimalapas still had good timber for masting, although it would be costlier to exploit. However, the felling campaigns undertaken between 1765 and 1772 were the last ones, because navy officers in Havana decided to begin sourcing their masting from Louisiana, and a supply centre was established in New Orleans for this purpose.

5   Forest Resources and Naval Wood Sourcing in Sierra Madre
       Oriental

In the second half of the 18th century, the strategic importance of New Spain grew gradually, and the need to reorganise its defences mobilised the viceroys; it was no longer a question of seeing off the sporadic attacks of pirates, corsairs, and buccaneers, but of holding up the advance of Great Britain in their strategy to wrestle colonial control from the Spanish. The British colonies in North America revolted while Bucareli was viceroy in New Spain; the Spanish Crown helped the rebels with weapons and money, while the courts of Madrid and Paris negotiated their intervention to expel the British from Florida and the Gulf of Mexico, and even to destroy all their bases in the Mosquito Coast and invade Jamaica. In addition to this, a new actor appeared on the competitive American colonial stage, with Catherine II's Russia looming over Alaska. The old territorial disputes with Portugal, which went back to the treaty signed in 1750, came to a head; in 1776 a Portuguese expedition attacked Colonia de Sacramento, triggering a Spanish response led by Pedro de Cevallos. These circumstances were a powerful incentive to fortify the territory, reorganise the army and the militias, and intensify the navy's presence in American waters.[47]

These threats made the Crown keep a particularly vigilant eye on the viceroyalty of New Spain, where administrative, fiscal, economic, commercial, and, it goes without saying, military reforms were launched. One of the first officials to come directly to grips with the situation was José Bernardo de Gálvez y Gallardo, 1st Marquis of Sonora, appointed Visitador General in New Spain in early 1765. His mission was both to ensure that Charles III's reform programmes were strictly implemented and to compile enough information to improve fiscality in the viceroyalty and promote the royal monopoly over the tobacco

---

47   José Antonio Calderón Quijano (ed.), *Los virreyes de Nueva España en el reinado de Carlos III* (Seville: Escuela Gráfica Salesiana, 1967); María Luisa Ramos Catalina y de Bardaxí, "Expediciones Científicas a California en el siglo XVIII", *Anuario de Estudios Americanos* 13 (1956), 217–310.

trade. He was also invested with powers to investigate the performance of Viceroy Joaquín Juan de Montserrat y Cruillas (1760–1766), against whom many complaints had been filed at the royal court. Finally, open conflict broke out between Gálvez and the viceroy, who was dismissed from his post in August 1766.[48] The Visitador General was in New Spain for five years, during which he became intimately acquainted with the administration of the territory, taxing, the operation of the *Real Hacienda*, mining, the transport of mercury, and the exploitation of salt mines; he also investigated contraband activities, a major problem, and the issues that undermined legal trade. Before he left for Spain he handed a detailed report to the new viceroy, Bucareli: *Informe instructivo del Visitador General de Nueva España al Excmo. Sr. Virrey de ella D. Antonio Bucareli y Ursúa en cumplimiento de la Real Orden de 24 de mayo de 1771*. Although this report does not refer to naval construction, it mentions the wood available in several regions of New Spain. Gálvez pointed out the rich woodland between Orizaba and Córdoba, whose resources could be used for various purposes. The Visitador General went into greater detail concerning the department of San Blas: he wrote that excellent woods were to be found near the harbour for the construction of "long and hardy ships",[49] that two packet-boats had been built there, and that one frigate commissioned by the Marquis of Croix was almost finished.[50]

The new viceroy, another important figure in the administrative, economic, and military development of New Spain, tried to act upon several of the observations contained in Gálvez's report, for instance, with the reform of the militia in several regions of the viceroyalty. Croix had already tried to revitalise naval construction, bringing about timber extraction for naval masting in the hills of Chimalapas, which was still ongoing in the final year of the Marquis of Cruillas's tenure (1765),[51] and granting the two first private *asientos* to supply the Armada with timber. The first, in 1766, was awarded to Antonio Basilio Berdeja, acting on behalf of his father, Captain Don Andrés Berdeja; the five-year contract contemplated the supply of prepared parts of cedar and mahogany to the royal shipyards in Veracruz and Havana. The second was awarded two years later, when Viceroy Croix signed a contract with Domingo Ramón Balcarsel, who committed to supply the masting required for three 80-gun

---

48   Vicente Salvador y Monserrat, *El marqués de Cruilles, Biografía del Excmo. Sr. Teniente general D. Joaquín Monserrat y Cruilles, Marqués de Cruilles, Virrey de Nueva España de 1760 a 1766* (Valencia: Nicasio Rius, 1880).
49   "de mucha manga y de igual resistencia".
50   Archivo General de Indias (AGI), Estado leg. 34 doc. 35.
51   This will be explained in more detail in the next chapter.

ships, three 70-gun ships, four 60-gun ships, and six 30-gun frigates.[52] It is important to point out that the felling of pine trees in Chimalapas was under the direct supervision of the *Real Hacienda* of New Spain and the provincial administration, specifically under the navy officials in Veracruz and the *alcaldes* of Tehuantepec, Guamelula, and Nexapa.[53] Antonio Basilio Berdeja's and Domingo Ramón Balcarsel's *asientos* were entirely private operations (both were members of the local elite of Veracruz).

As a result of increasing felling activity to supply the naval industry, in a letter dated 23 October 1766, the Marquis of Croix suggested creating a regular warehouse for masts and dressed cedar and mahogany parts in Veracruz, "for the Armada's urgent needs and [also] to sell to private merchants".[54] In addition, the Marquis also wrote that more forests could be found in the region of Veracruz, for instance in the Tampico hills; however, no surveys had been undertaken to explore their quality, so he recommended operations to continue in known regions, such as Chimalapas and Coatzacoalcos.

Viceroy Antonio María de Bucareli y Ursúa (1771–1779) was an active reformer. Among other things, he spent his time in office putting the *Real Hacienda* back on a sound footing and reactivating mining operations, while reinforcing the three imperial frontiers of his viceroyalty: the Atlantic coast (the so-called Seno Mexicano), the Pacific coast, and the northern frontier. He improved the fortifications of San Juan de Ulúa, Perrote, and Acapulco, and designed a network of presidios in northern New Spain to protect Spanish colonists in Sonora and Louisiana from the Apache and other native groups. When he heard that a Russian expedition, led by Lieutenant Tschericow, had explored the western coast of North America between 1769 and 1771, Bucareli sponsored three Spanish expeditions to the Pacific coast north of Mexico, in 1774, 1775, and 1779. These expeditions revealed that no foreign settlement had been established, and that the strait of Anián, the alleged northern pass linking the Pacific and Hudson's Bay, did not exist. This was good news for the Spanish, who had feared a British attack from that side.[55]

Concerning the exploitation of woodland, Bucareli was the royal official who expressed the greatest interest in surveying the badly known woodlands of Sierra Madre Oriental. He promoted several survey projects aimed at finding

---

52   Reichert, "Recursos forestales, proyectos de extracción y asientos de maderas", 71–72.

53   AGNM, Industria y Comercio, vol. 10, exp. 1, fs. 1–47, Corte real de arboladuras en Chimalapa.

54   "las urgencias de los bajeles de Armada y [también] para vender[los] a los mercantes".
      AGNM, Correspondencia de Virreyes, serie 2, vol. 11 exp. 57 f. 102.

55   Ramos Catalina y de Bardaxí, "Expediciones científicas a California en el siglo XVIII",
      217–310.

hardwoods that could be used for naval construction. These expeditions covered from Veracruz[56] to Laguna de Términos. His original proposals to explore the woodland in Teziutlán and Perote, presented in 1774, were endorsed by a royal order dated 28 February 1776 in which Charles III expressed the need to "increase my fleets, in both Spain and the Indies, as much as possible".[57]

6 Antonio María de Bucareli y Ursúa's Land Survey Instructions

With the support of this royal order, in July 1776 Bucareli issued a report suggesting the resumption of shipbuilding in Tlacotalpan, in the Alvarado River, where "some warships, like the *Bizarra* and the *Nueva España*" had been built in the past.[58] This is not entirely correct, because the former was in fact built in Havana and the latter in the bar of Coatzacoalcos. This is not to say that Tlacotalpan was new to shipbuilding; two merchant ships, the *Gallo Indiano* (425 tons), launched in 1722, and the *Paloma Indiana* (643 tons), launched in 1725, had been built there. After serving as merchant ships for a short time, both were bought by the *Marina Real* and refitted as warships, with 58 and 52 guns, respectively.[59] The record also mentions the construction of a third ship in Tlacotalpan, the *Rosario de Murguía*. All three ships proved to be sound vessels, despite their comparatively low cost.[60]

Interestingly, the king made Bucareli negotiate with the Church councils of New Spain, the Mexican consulate, and other commercial and mining institutions to fund the construction in Tlacotalpan-Alvarado of 50- to 70-gun ships and 30- to 40-gun frigates.[61] Spanish monarchs often involved merchants, miners, and the Church in their projects, especially on the eve of war or during wartime. These subsidies to the Crown were referred to as *donativos*. This practice is illustrated by the construction of ships-of-the-line in Havana during the American War of Independence and, after the war, in the 1780s. Another well-known example is the construction of the 112-gun *Nuestra Señora de Regla* or

---

56 The region between the Alvarado and Coatzacoalcos rivers.
57 "aumentar, cuanto sea posible, sus armadas navales así en España como en las Indias". AGNM, Correspondencia de Virreyes, serie 2, vol. 12, exp. 159, f. 26.
58 "algunos bajeles de guerra, entre ellos la *Bizarra* y la *Nueva España*" (Ibid.).
59 Gervasio de Artíñano y Galdácano, *La arquitectura naval española (en madera)* (Madrid: Oliva de Vilanova, 1920), 57–58; Bibiano Torres Ramírez, *La Armada de Barlovento* (Seville: Escuela de Estudios Hispano-Americanos, 1981), 96.
60 AGNM, Correspondencia de Virreyes, serie 2, vol. 12, exp. 162, f. 31v.
61 Ibid., exp. 159, f. 27.

Conde de Regla, financed by Pedro Romero de Terreros, one of the leading merchants and mining entrepreneurs in New Spain, in 1786.[62]

The viceroy sent engineer Miguel del Corral and a navy officer from the San Blas department to find the best location for the shipyard. Their orders said: "survey the ground, the forests, the proportion of different woods, ease of construction, supplies, the depth of the river and the bar, and the possibility of dredging it to open the way to the ships".[63] The viceroy put a ship and an experienced crew at the disposal of the surveyors to sound out the river and also promised carpenters to examine the wood. All of this, as Bucareli stressed, contributed to increasing the naval forces, "which are the main support for the conservation of these [American] dominions".[64] In another report, dated 27 August 1776, the viceroy gave further information about the expedition to the Alvarado River. Interestingly, by the time this report was issued the range of the survey had been increased with the inclusion of the Coatzacoalcos River. The old provisional shipyards in both rivers were to be explored to determine their suitability and define the best location for the new ones. A first visit to the region around the Coatzacoalcos emphasised the abundance, and accessibility, of wood in the surrounding forest.[65] The Alvarado, however, was barely twelve leagues from Veracruz, presented better mooring conditions, and the bar was better suited to build a fortification. The nearby woodland also offered ideal forests for naval construction. The viceroy also pointed out that there were many skilled ship carpenters in Tlacotalpan, and that they had been involved in felling and dressing trees for ships of various sizes since the early 18th century.

The exploration of the territories of Alvarado, Tlacotalpan, and Coatzacoalcos began right away, as engineer Miguel del Corral and Frigate Captain Joaquín de Aranda left Veracruz in October 1776. Captain de Aranda, instead of an officer from the San Blas department, was chosen for the mission by the fleet commander, Antonio de Ulloa. Bucareli was trying to mobilize the chief

---

62   Guillermina del Valle Pavón, *Donativos, préstamos y privilegios: los mercaderes y mineros de la Ciudad de México durante la guerra anglo-española de 1779–1783* (Mexico: Instituto Mora, 2016: 44).

63   "reconocer el terreno, los bosques, proporción de maderas, facilidad de las obras necesarias, subsistencia de víveres, fondo del río y su barra, y disposiciones de mejorarla para dar salida a los buques que puedan construirse allá". AGNM, Correspondencia de Virreyes, serie 2, vol. 12, exp. 159, f. 27.

64   "deben ser el principal apoyo de la conservación de estos dominios americanos". AGNM, Correspondencia de Virreyes, serie 2, vol. 12, exp. 159, f. 27v.

65   AGNM, Correspondencia de Virreyes, serie 2, vol. 12, exp. 162, f. 30.

of New Spain's latest fleet to help him build a shipyard in the jurisdiction of Veracruz,[66] and the details of the officer's mission, including his task of finding a good location for a shipyard in Coatzacoalcos or in the village of Tlacotalpan, were mentioned in the *Relación* published in 1777.

7 Miguel del Corral and His Contribution to the Scientific Knowledge of New Spain

Military engineer Miguel del Corral arrived in Veracruz in 1764, aboard the expedition organised by Brigadier Juan de Villalba to create the army of New Spain. In 1766, Corral embarked on an expedition to recognise the coast of the Gulf of Mexico, about which he was to write two memorandas, which were the foundation of a series of works about the geography of Veracruz. In 1767, he drew a map entitled *Mapa de la porción de las Costas del Seno Mexicano, comprendida entre la Barra de Alvarado y Cabo Puntillas*. In 1768, he was in Mexico City working in the construction of a lime and brick factory, and two years later he was in Perote, building a fort. In 1771, Viceroy Bucareli commissioned engineers Agustín Crame and Miguel del Corral to carry out the preliminary studies for the construction of a fluvial connection between the Atlantic and the Pacific oceans. The idea was to establish whether the Coatzacoalcos River would be suitable for the construction of an inter-oceanic channel through the isthmus of Tehuantepec. In 1773, Miguel del Corral, by then a colonel, was drawing plans to leeward and windward of Veracruz along with another six engineers. In 1775 he was sent to the Pacific harbours of San Blas, Matachel and Chacala. He drew *Plano y mapa reducido de un cuadrado fortificado que se proyecta para defender la entrada de la barra y río Coatzacoalcos* (see Figure 9), dated to 12 October 1778, and also *Planos y perfiles de un fuerte para defender la entrada de la barra y río de Alvarado* (see Figure 10); in 1779 he drew *Mapa de la Costa de Veracruz, desde Alvarado hasta Boquilla de Piedras*; and, finally, in 1780, along with engineer Manuel Santisteban, *Elevación y perfil de un puente sobre el río la Antigua*. In 1780, del Corral was appointed chief engineer, and on 19 March 1781 Lieutenant commander of the Castle of San Juan de Ulúa. In 1782, Miguel del Corral, as the king's lieutenant in San Juan de Ulúa, issued, in response to Viceroy Matías de Gálvez's petition, his *Relación circunstanciada del estado de las fortificaciones y edificios militares en la Plaza de Veracruz, su*

---

66   Ibid., exp. 162, f. 34.

costa y castillo de San Juan de Ulúa, in which he gave his negative opinion about the fortification. In 1784, del Corral drew a project for the water supply of Veracruz, including a general map and the course of a water channel from the Jamapa River. Also in 1784, he drew up the plans for the customs building in San Juan de Ulúa. Until his death in 1794, del Corral remained very active. His career earned him the highest rank in the Engineer Corps: in 1783, he was put in charge of the Dirección de Ingenieros de la Nueva España, and in 1790 he was promoted to brigadier. In addition, he was temporary governor of Veracruz (1782–1786), Lieutenant of the Fortification of Veracruz and Real Fuerza de San Juan de Ulúa (1783), and intendent of Veracruz (1790–1793).[67]

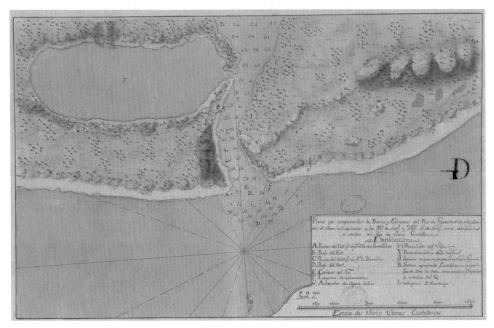

FIGURE 9   Plan that includes the bar and entrance of the Coatzacoalcos River from the Gulf of Mexico
SOURCE: ESPAÑA. MINISTERIO DE CULTURA Y DEPORTE. ARCHIVO GENERAL DE INDIAS,
MP-MEXICO, 331 BIS

---

67   Horacio Capel Sáez, *Los ingenieros militares en España. Siglo XVIII: repertorio biográfico e inventario de su labor científica y espacial* (Barcelona: Universitat Barcelona, 1983);
José Antonio Calderón Quijano, *Historia de las fortificaciones en Nueva España* (Sevilla:
Escuela de Estudios Hispano-Americanos, 1953); Omar Moncada Maya, *Ingenieros militares en Nueva España. Inventario de su labor científica espacial. Siglos XVI a XVIII* (México: Universidad Nacional Autónoma de México, 1993).

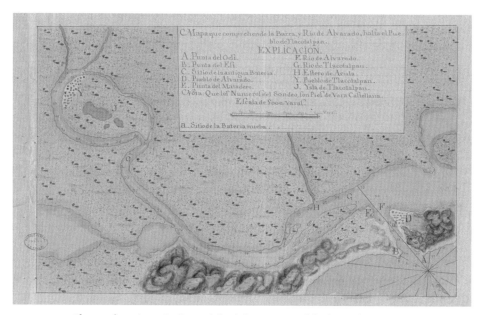

FIGURE 10   Plans and sections of a fort to defend the entrance of the bar and river of Alvarado and
Tlacotalpan village
SOURCE: ESPAÑA. MINISTERIO DE CULTURA Y DEPORTE. ARCHIVO GENERAL DE
INDIAS, MP-MEXICO, 385 TER

8   Relation of the Survey Carried out in the Hills of Teziutlán and
Perote (Veracruz)

The earliest survey of the hills and woodland of Sierra Madre Oriental was
undertaken by engineer del Corral in May 1774, when he was appointed by the
head of the Real Maestranza de Veracruz, Joseph Segurola, to assess "masting
for all kinds of ships, and their transport to this city [of Veracruz] from the hills
of Teziutlán and Perote, where it is said there are suitable pine trees for this
purpose".[68] The general administrator of the *Real Hacienda*, Bernardo Nicolás
de Lemos, added that engineer Corral's mission aimed to find trees for masting

68   "arboladuras, para toda clase de embarcaciones y su conducción a esta ciudad [de Vera-
cruz] desde los montes de Teziutlán y Perote, donde según dicen hay pinos apropósito
para el efecto". AGNM, Industria y Comercio, vol. 10, exp. 5, fs. 101–111v, Relación del reconocimiento hecho en los montes de Teziutlán, y Perote por el Teniente Coronel de Ingenieros Don Miguel del Corral, en solicitud de maderas aptas para toda Arboladura de
Navío.

near Veracruz, because those from Coatzacoalcos were too costly, owing to transport expenses from the forests of Chimalapas. The Report includes a detailed analysis of hills, terrain, and rivers in several areas, for instance, Teco-lotepeque, which was 6 leagues from Perote, where there was a substantial number of Mexican white pines, which the locals called *ayacahuite*. However, the distance to the Santa María Tlapacoyan River, which runs into the bar of Nautla, in the Seno Mexicano, ruled out this option because of the cost of transport and of preparing the paths; the river was at the bottom of steep slopes and narrow passes. The engineer also visited the woods of Macacalco, 10 leagues from Perote, 12 from Jalapa, and 34 from Veracruz. There he found an abundance of good-quality pines of all sizes, but distance, and the rivers that ran across the way, remained a problem; the path that came down to a place that the locals called La Pila ran across broken terrain and would need to be prepared for the timber to be carried to the bar of Nautla. The cost was again considered too high.[69] Del Corral indicated that the places of Atopa, Llano del Paisano, Potrero de los Caballos, and Alto de Villegas, in the region of Peña del Cofre, were equally rich in pines of all sizes, although not in such abundance as in Mecacalco, because many had been felled to meet the needs of the city of Jalapa.[70]

Del Corral found time to survey the woodland mass south of Peña del Cofre, 5 leagues from Perote and 37 from Veracruz, where he found a large mountain covered in many pines of all sizes and very clean, that is, without dry branches and bark beetles. The officer observed that the Potrero de Ygnes María and a place called Chololoyan were suitable to begin felling and that the timber could be brought through the village of Ysguacan de los Reyes,[71] more or less 5 leagues downslope. This route was formerly used by the locals to drag timber with oxen. Three leagues, also going downhill, separated Ysguacan from the Güisilapam River;[72] the most abrupt passes could be travelled on horseback. Near this tributary river, there was a place called Teya, from which timber had been sourced in the past. Del Corral pointed out that, in order to obtain full masts,[73] the road would need to be prepared, so that the oxen could be used all the way. Asking local fishermen, he learned that rafts could be used downriver all the way to La Antigua when the water ran high, without natural obstacles, rapids, or narrow passes. However, del Corral was aware of the need to know

---

69   AGNM, Industria y Comercio, vol. 10, exp. 5, fs. 102v–103.

70   Ibid., f. 103v.

71   Modern Ixhuacán de los Reyes.

72   The Chichiquila River, which runs into the Antigua River.

73   A frigate or a ship-of-the-line requited approximately 36 masts of different diameters.

the river in detail, writing this "although in many parts it can be walked on foot, the fishermen rafts cannot pass, whatever they say".[74] He also said that, although the slopes of Peña del Cofre were covered in trees, these were hard to access; the area was rich in other tree species too, such as oak, holm oak, and elm.[75]

Interestingly, del Corral estimated the cost of the felling, dragging, and transport operations between Ysguacan de los Reyes and the mouth of the Antigua River in the Gulf of Mexico. According to his calculations, the costliest part would be to bring the masts from the felling area to the Güisilapam River, which would require a large workforce (1,000 labourers), at a total cost of 6,250 pesos de a ocho reales; the felling and dressing of the trunks, in contrast, would barely fetch 360 pesos, and the same amount for the transport downriver to La Antigua (see Table 42).

His work *Relación del reconocimiento hecho en los montes de Teziutlán, y Perote por el Teniente Coronel de Ingenieros Don Miguel del Corral, en solicitud de maderas aptas para toda Arboladura de Navío*, in which he gave detailed descriptions of the wood and the obstacles that could get in the way of felling and both overland and downriver transport operations, illustrates Miguel

TABLE 42   Felling, dragging, and transport costs for one full masting from the hills of Ysguacan de los Reyes to La Antigua

To fell and dress the 36 main masts, 10 pesos each	360 pesos
To clean the hilly area towards the royal road and cut a dragging area	600 pesos
To take them out of the hills, and into the road	400 pesos
To drive them from the road to the riverside, through the longest route, estimating 1,000 men for 20 days, at a rate of 2.5 reales per day	6,250 pesos
Bringing rafts to La Antigua, estimating five rafts, each with six bargemen, at a rate of 12 pesos for the trip	360 pesos
For unexpected expenses, and other small details involved in the preparation of masting	2,000 pesos
For preparing the road from the logging area to the Güisilapam River, changing its course in places, and improving it as much as possible	2,500 pesos

SOURCE: AGNM, INDUSTRIA Y COMERCIO, VOL. 10, EXP. 5, F. 105

---

74   "aunque en muchas partes se anda, no pueden cruzarlas balsitas por contrario que sirven a los pescadores". AGNM, Industria y Comercio, vol. 10, exp. 5, f. 104v.

75   Ibid.

del Corral's excellent scientific credentials. In his following expedition, to the region between the Alvarado River, in Chimalapas, and the Coatzacoalcos River, he again proved to be equal to this sort of major mission.

9   The Major Survey Expedition of Miguel del Corral and Joaquín de
       Aranda to South of Veracruz

The *Relación*[76] that Miguel del Corral wrote during his expedition, between October 1776 and July 1777, is divided into five parts. The first is a geographical description of the ground surveyed alongside the navy pilot Joaquín de Aranda, and it includes a map (see Figure 11); in the second, del Corral described the hills on which he found wood suitable for naval construction and their condition; in the third, he examines the possibility of building a shipyard in the bar of Coatzacoalcos or the village of Tlacotalpan; in the fourth del Corral mentions the fortifications in the mouths of the Alvarado and Coatzacoalcos rivers; and the fifth addresses the militias that he found in some of the villages during his survey. For this book, the second section in particular, with a detailed account of the species of wood suitable for naval construction available between Tlacotalpan and Coatzacoalcos, has been of great use.[77]

The first area whose wood resources were described in the *Relación* published in 1777 was that between the Tlacotalpan and San Juan rivers: "ordinary cedar, mahogany, male cedar, tavi, zapote, guayacán, palo maría, cocuite, and holm oak are driven down all the streams, to both the right and the left".[78] The description, therefore, claims that timber had long been sourced from the area, for domestic purposes but also for naval construction. According to Germán Andrade, the area began supplying timber for private ships and to the shipyards in Veracruz, Havana, and La Carraca in the early 18th century.[79]

---

76   Full title: "Relación de los reconocimientos practicados por el coronel ingeniero en
      segundo don Miguel del Corral y el capitán de fragata graduado y piloto mayor de der-
      roteros de la Real Armada don Joaquín de Aranda, de orden del excelentísimo señor don
      Antonio María Bucareli y Ursúa, Virrey Gobernador y Capitán General del Reino de la
      Nueva España; a que dieron principio por la barra de Alvarado en 28 de octubre del año
      pasado de 1776 y finalizaron el día de la fecha [en Tlacotalpan el 21 de julio de 1777]".
77   Siemens and Brinckmann "El sur de Veracruz a finales del siglo XVIII", 311–316.
78   "por todos los arroyos de la derecha e izquierda, empezando desde la hacienda del
      Zapotal, se han sacado maderas de cedro ordinario, caobano o cedro macho, tavi, zapote,
      guayacán, palo maría, cocuite, y encino".
79   Andrade Muñoz, *Un mar de intereses*, 87–98.

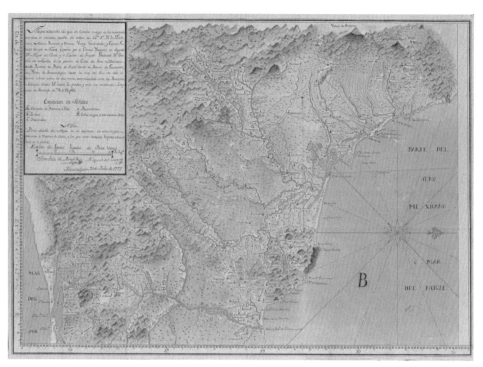

FIGURE 11   Map of the coast of the Mexican Sound from Puntilla de Piedra to Coatzacoalcos, which
also includes the Isthmus of Tehuantepec (1777)
SOURCE: ESPAÑA. MINISTERIO DE CULTURA Y DEPORTE. ARCHIVO GENERAL DE
INDIAS, MP-MEXICO, 329 BIS

Del Corral also pointed out that previous felling had left some clearings in the woods, but the exploitation of this area of woodland continued as far as the pass of San Juan. He also made an interesting observation: the *haciendas* of Nopala, Solcuautla, and Cuatotolapan had good mahogany, but the owner of these estates, Don Diego Fentanez, had forbidden them to be felled. From 1748, this landowner was in open conflict with wood contractors, who demanded access to his estates. It is to be noted that his first antagonist was Don Andrés Berdeja, one of the main contractors for the Armada de Barlovento and a highly influential figure in the elite of Veracruz, who probably also represented the interests of the merchants Consulado de México in the province.[80] Del

---

80   Reichert, "¿Cómo España trató de recuperar su poderío naval?", 83.

Corral mentions this contractor in one paragraph of his *Relación*: "Captain Don Andrés Berdeja, who is steadfast, a man of spirit and knowledge in the felling business, was based in the Zapotitlán River for over two months [...] to fell some pines, but left after his hopes were shattered".[81]

The hills of Nopala, Solcuautla, and Cuatotolapan, which had caused so much acrimony between Fentanez and Berdeja, were described by del Corral as rich in cedar, "of thick, long, and high-quality wood",[82] also mentioning that the place was well communicated with the Tesechoacán and San Juan rivers, which would make it easy to drag the trees to the rivers using oxen. In the next step of his survey, Miguel del Corral visited the *hacienda* of Santa Catalina de los Pozos, where he spotted much ordinary and male cedar, which would also be easy to bring to the Gulf coast through a spring that was 5 leagues distant from the forest. Afterwards, del Corral followed the course of the Candallón River, where he identified ordinary cedar and some male cedar. There, he also found three active felling areas, one near the *hacienda* of Santa Catalina, another one around the Tomás Martín Stream, and a third one higher up the hills. He observed that, in the second one, a group of workers was carving canoes, and that four of them were nearly complete. He also spotted 20 large, felled trunks of mahogany, which were to be carved into canoes too, and that paths into the jungle had been opened to bring the timber out.[83]

After this, del Corral headed to the region of the Tataguicapa, La Lana, and Sanguluacan rivers, where he pointed out the presence of abundant cedar in some areas, like the Hill of Coyole, that belonged to indigenous villages; these did not consent to the felling of their trees. In contrast, the areas of Lomba, Corral Nuevo, and Laguna Redonda, in the hill range of Tuxtla, also had abundant wood, the felling of which had begun soon earlier. Despite the intensity of the ongoing operations, there were still trees on offer, but they would be costly to exploit, as some grew in hard-to-access terrain. Cedar could also be found around the Tesechoacán River, which joined the Cosamaloapan. Del Corral reported four active felling areas in the region: two near the Caxonos Stream, one near the Manso River, and one near the Chilapa River, in the jurisdiction of Guaspaltepeque. Similarly, he pointed out that the hillsides were covered in trees, although these were far from the active felling areas, which sought the

---

81   "el capitán don Andrés Berdeja, que es constante, fue un hombre de mucho espíritu y práctica para cortes de madera, estuvo establecido en el río de Zapotitlán más de dos meses [...] para sacar pinos, y se retiró sin esperanza de conseguirlo". Siemens and Brinckmann, "El sur de Veracruz a finales del siglo XVIII", 313.

82   "madera muy gruesa, larga y de buena calidad".

83   Siemens and Brinckmann, "El sur de Veracruz a finales del siglo XVIII", 311–312.

proximity of water courses. Still in this region, he inspected the hill of Mirador and the surrounding elevations, part of the estate of Chiltepeque, near the Poposoca River, where he found large expanses of virgin cedar woods.[84]

Concerning the Tesechoacán River, it is interesting to note that one of the earliest felling areas was established following the award of an *asiento* to Don Vicente Villa in 1756–1757; this led to the construction of a mooring area for timber in one of the river's beaches, from which they were sent to Cosamaloapan, Tlacotalpan, Alvarado, and probably Veracruz, for naval construction. It is also worth pointing out that in these contracts, the main *asentista* subcontracted the work to two, three, or more woodmen, who spent months at a time in their allocated areas. The subcontractors hired local workmen—in the case of this contract, from Chilapa—but also their own hands, who specialised in felling, dragging, and rafting timber. In this instance, Villa hired mixed-race lumberjacks from Tlacotalpan.[85]

Following this, del Corral entered the jurisdiction of Tuxtepeque, where he found good cedar wood and other useful species for naval construction near the banks of the Chicintla River, an area seldom frequented by woodmen. He also found abundant cedar near the Santo Domingo River, where wood had not previously been sourced intensively, as it was only exploited by locals to build and repair their houses and churches. Finally, del Corral pointed out that, in the area between both banks of the mouth of the Tonto River and the pass of Soyaltepeque, there was wood in abundance, and that the nearby hillsides were also rich in cedar. According to his report, there were enough trees in the area to last for many felling seasons. He also spotted other hardwoods, such as mahogany, in various places, although he clarified that some of the groves were richer than others. Finally, del Corral reported a holm oak grove on the right bank of the Tonto River, going from Pueblo Nuevo to the pass of Soyaltepeque (modern Soyaltepec), and another one, which was also fairly rich, from the pass of San Juan to Acayucan. He described both of these mahogany groves as highly valuable.[86]

When he was done inspecting these areas, del Corral headed to regions rich in pine, which was badly needed for naval masting. First, he arrived at Zapotitlán and Tepetotuptla (modern Tepetotutla), where trees were abundant. The

---

84   Ibid., 312.
85   Rafael Palma and Odile Hoffmann "La conformación de una frontera interna en las riberas del Tesechoacán", in María Teresa Rodríguez and Bernard Tallet (eds.), *Historias de hombres y tierras. Una lectura sobre la conformación territorial del municipio de Playa Vicente, Veracruz* (México: Centro de Investigaciones y Estudios Superiores en Antropología Social, 2009), 53.
86   Siemens and Brinckmann, "El sur de Veracruz a finales del siglo XVIII", 313.

terrain, however, was difficult, and the rivers were full of strong currents and many rocks. As such, unless something was done about it, taking the tree trunks downriver to the Gulf of Mexico would not be possible. The experience of Antonio Berdeja, son of Don Andrés, proved this point; after felling some trees in this area, he found that he could not take them downstream, and was only able to bring a few pieces to the village of Usila.[87]

Back to the 1777 *Relación*, del Corral also mentions other areas between the Alvarado and Coatzacoalcos rivers, and the hills towards Tehuantepec and Sierra Madre Oriental, where pine, ideally suited for masting, abounded. Del Corral notes that he heard this from Don Lorenzo Arrinda, who "as a lieutenant in the Crown's regiment knew something about the felling of trees for mast-ing, after spending many years in the felling areas near Coatzacoalcos".[88] Simi-larly, del Corral held Arrinda's detailed knowledge of the hills in the region of Zapotitlán and Tepetotuptla in great consideration; Arrinda had calculated that between 30,000 and 40,000 pesos de a ocho reales were needed to over-come the obstacles posed by the rivers, so del Corral argued that it was not worth opening new paths in the area to transport the timber, because of the cost.[89]

Continuing his survey, del Corral reached the Coatzacoalcos River, where there was much ordinary cedar, male cedar, guapinole, oak, palo maría, and holm oak. He entered the area between the streams of Chacalapa and Guasun-tan, down which the timber used in the ship *Nueva España* had been driven. He pointed out that, on the hillsides of the range of Minzapam, there was still an abundance of timber that needed to travel only a relatively short distance. It may be interesting to recall the story of this vessel, built as part of the attempt to create a shipyard in Coatzacoalcos and develop the naval industry of New Spain. The project took off with the idea of building two 64-guns ships-of-the-line, but the reluctance of navy officers in Havana to send shipwrights forced the project's promoters to downsize it to a single ship, called *San José*, alias *"Nueva España"*, with a 66-cubit keel and carrying 60 guns. As previously noted, the construction spanned 1731 and 1734 and cost a total of 431,000 pesos de a ocho reales, footed by the *Real Hacienda* of New Spain. The cost was so high because of the mortality rate among the woodmen, as the felling areas were situated in an unwholesome area teeming with *moyocuile* mosquitos. In the

---

87   AGNM, Industria y Comercio, vol. 10, exp. 2, fs. 50–63v, Asiento de maderas celebrado con
      Antonio Berdeja, apoderado del capitán Andrés Berdeja.
88   "siendo teniente del regimiento de la corona tenía inteligencia en los cortes de arboladura
      por haber estado muchos años en los cortes de la región de Coatzacoalcos".
89   Siemens and Brinckmann, "El sur de Veracruz a finales del siglo XVIII", 313.

event, the enormous administrative and logistic problems faced by the project, corruption, and the high cost of labour, food, and tools, prompted the viceregal authorities to cancel the projected shipyard and continue building their ships in Havana.[90]

Del Corral then set off for the head of the San Antonio River, in the jurisdictions of Isguatlan, Moluacan, and *hacienda* San Antonio, where he found areas of virgin woodland of male cedar and palo maría. On reaching the Coatzacoalcos he continued to Malpaso, where he noted the presence of ordinary cedar and male cedar. According to his observations, ordinary cedar was less abundant and more dispersed, but male cedar was found in much greater quantity; del Corral described a specific grove, approximately one square league in size, with between 800 and 1,000 male cedars in it. He made a similar discovery on the banks of the Los Mijes River, where both species were also to be found. Throughout the region, del Corral also spotted other hardwoods, including zapote, oak, guayacan, and palo maría, interspersed among the cedars. Del Corral ended this section of the *Relación* with an important fact: he mentions that the Coatzacoalcos River and its tributaries were seldom used to transport wood, and only for domestic purposes, and that the only large-scale felling to have taken place in the area was that related to the building of the *Nueva España*, as well as a cargo of large trunks felled near the Los Mijes River, which had been sent by barge to Tlacotalpan.[91] It is clear that del Corral saw great potential in the region between Alvarado, Coatzacoalcos, and Sierra Madre Oriental, a potential that was to go untapped by the Crown's naval departments. In this, the shortage of vessels to transport the timber, difficulties with the felling seasons, arrears in pay that often ground work to a halt, the robbing out of tools, the loss of timber to accidents with river rocks, and the impossibility to float the trunks when the river waters were running low, also played a part in the forests of Veracruz remaining underexploited.

10 Information about the Exploitation of Natural Resources in Miguel del Corral's *Relación*

At the time when Miguel del Corral was surveying the hills to south of Veracruz, there was no legislation to regulate the exploitation of wood resources in continental New Spain. In Spain, this activity was regulated by the *Real Ordenanza sobre Bosques y Plantíos*, enacted in 1748. According to del Corral, this

---

90   Andrade Muñoz, *Un mar de intereses*, 85–87.
91   Siemens and Brinckmann, "El sur de Veracruz a finales del siglo XVIII", 315.

lack of rules concerning the felling and dimensions of New Spanish timber led
"everyone to cut wood as they please [...] without any attention to size, wiping
everything out, including saplings, which are barely good enough for a square
sesma, which ruins the best-curved woods [for naval construction], wasting
much to take straight masts ten varas long".[92] Del Corral, who was apparently
aware of the 1748 *Ordenanza*, mentions that there was no interest in planting new trees, as established by the metropolitan regulations; he knew of no
planting of cedar or other species suitable for naval construction. At the same
time, he denounced that the felled trunks were piled up on the riverbanks to
float them during the rainy seasons, making them dry out and lose their value
for naval construction; these losses, in turn, led to uncontrolled clandestine
felling over increasingly large areas and influenced the wood-sourcing strategies followed by locals, who destroyed many trees good for beams and slimmer
masts.[93]

In order to solve this, del Corral suggested regulating felling and impos-
ing legal punishments for non-compliance, for instance, the cutting down of
cedars from which less than half a vara of board could be got. This was supported by his observations in riverbanks and nearby areas, where young cedars
were found in abundance; he also recommended planting large areas of new
trees every year, so that sufficient timber for all the king's ships to be of good
cedar was always available. Miguel del Corral also observed that, in the lati-
tudes that he was surveying, cedar grew faster than in other northern regions.
He expressed surprise at the incredulity of local ranchers and farmers, who
claimed that cedar could not be planted on riverbanks, despite the abundance
of saplings there, no doubt the issue of waterborne and windborne seeds. In
this and similar passages, del Corral betrays an excellent understanding of the
management of forest resources; he even outlined a project for the *hacienda*
owners and the local authorities to plant trees. His *Relación* recurrently mentions the fact that these woods were necessary for the royal service, and that
the king's arsenals needed to be supplied with good wood, be it through the
work of royal officials or by *asiento*.[94]

Another measure that del Corral recommended in his *Relación* was to forbid
private agents, including *hacienda* owners and the residents of the villages in

---

92   "cada uno corta, y ha cortado las maderas según su antojo o conveniencia [...] la han
       cortado sin atención ninguna a su tamaño, arrastrando con todo, hasta con los pimpollos
       que apenas daban una sesma en cuadro, dejando pérdida o extraviada la mejor madera
       como curvas muy apreciables [en la construcción naval], echando a perder mucha para
       sacar palos derechos de diez varas" (ibid.).

93   Ibid.

94   Ibid., 315–316.

whose land woods under royal protection grew, from exploiting them. Woods suitable for naval construction were to be marked, but this task could take years, during which the king would reap little profit from them. In order to save time, he suggested instructing *hacienda* owners and local authorities to mark the woodland themselves, which would also help prevent illegal felling, as each *hacendado* and *justicia real* were to present an inventory with the number of trees of each species found in their jurisdiction. Miguel del Corral was convinced that reserving cedar and other hardwoods for the royal factories would not be excessively detrimental to *indios, mestizos, pardos,* and Spaniards, because the uses they gave to wood could be perfectly met with other species, such as the *ocote* and the *zapote*. Licences to fell wood for the construction of houses, chapels, and churches would be issued, and the felling would be undertaken under the supervision of the local authorities. Del Corral also left open the possibility of granting private *asientos* to supply the royal shipyards.[95] The 1748 *Ordenanza* presented similar proposals to benefit the *Marina Real* and limit private felling, which sparked conflict between the owners of woodland and the Armada's intendents in charge of supervising it. Notably, the outcome of this situation was a reduction in supply for the naval departments of Cartagena, Cádiz-La Carraca, and El Ferrol, forcing the Secretary of the Navy to import Baltic, Italian, Romanian, and American wood.

11 The Survey of the Antigua and Nautla Rivers and the Mecacalco Hills

Back to the woodland surveys directed by engineer Miguel del Corral in Sierra Madre Oriental, the exploration of the Antigua and Nautla rivers was undertaken in August 1778, and of the Mecacalco hills in the following September. The survey projects were outlined by Colonel Don Miguel del Corral: "instructions for José Vicente, *vecino* of Tlacotalpan appointed by the Colonel to survey the pine groves of Antigua and Nautla, with the purpose of establishing if the masting requested by the Havana squadron commander can be easily driven".[96]

---

95   Ibid., 316.
96   "informe, y de instrucción de lo que deba ejecutar José Vicente vecino de Tlacotalpan propuesto por dicho Coronel para el reconocimiento de los pinos de la Antigua, y de Nautla que corresponde hacerse con el fin de ver la mayor facilidad de conducir por ellos las Arboladuras que tiene pedidas el Señor comandante general de la escuadra de la Habana". AGNM, Industria y Comercio, vol. 10, exp. 10, f. 151.

The first expedition to the Antigua and Nautla rivers followed the initiative of the commander of the New Spanish fleet, Antonio de Ulloa, who, in his return voyage to the Iberian Peninsula, brought with him samples of the pines that grew near the volcano Orizaba and Cofre de Perote.[97] During his stop in Havana, which lasted for nearly a month, between February and March 1778, the fleet commander brought some of these samples to be inspected by the head masters of the arsenal, who reported that "these masts are incomparably better than those of Coatzacoalcos, and considering the use that these [naval] forces can make of them [...] I thought it worth pointing it out to V. E. [Viceroy Bucareli] so that you can order the felling and transport of the masting specified in the following note".[98]

The commander of the Havana department, Juan Bautista Bonet, wrote again to Bucareli in March 1778 to request his support for an expedition to fell said masts and bring them to Havana via Veracruz. Letters exchanged by the viceroy, engineer del Corral, and the administrator of the *Real Hacienda* in Veracruz, Pedro Antonio de Cosio, discussed the best place to undertake the felling and the costs to be shouldered by the Royal Treasury. The project was finally accepted by Viceroy Bucareli in July 1778, and on 10 August the explorers left the harbour of Veracruz. Finally, the plan was to survey several areas of pine forest populated by large and good-quality trees. As noted by del Corral:

TABLE 43   Masting requested by the Havana arsenal

Number of masts	Length in cubits	Hands in diameter
6	45	9
5	43	8
5	40	7½
4	38	7
6	36	6
8	30	5

SOURCE: AGNM, INDUSTRIA Y COMERCIO, VOL. 10, EXP. 10, F. 154

97   The samples were taken from two different locations.
98   "incomparable ventajosa de estos [palos] a los de Coatzacoalcos y considerado que esta escuadra, y demás fuerzas [navales] que se la puedan agregar y utilizar [...] me ha parecido importante hacerlo presente a V. E. [virrey Bucareli] para que se sirva disponer el corte, y conducción del número de perchas según las dimensiones que expresa la nota adjunta". AGNM, Industria y Comercio, vol. 10, exp. 10, f. 152.

"the most suitable in terms of size are those in the places of Chololoya and Zapotitlán; the former must come out down the Antigua River, and the latter down the Santa María Tlapacoya River and the bar of Nautla; the tree cover in these areas is constant throughout".[99] The mission was also entrusted with examining timber for masting in the place of Mecacalco, in the jurisdiction of Xalapa, near Santa María Tlapacoya,[100] and confirming whether the timber requested by the fleet commander in Havana, Bonet, could be floated down. The bill for the expedition, which included Miguel del Corral, his assistants José Vicente Martínez and Cayetano Rangel, and four woodmen from Tlacotalpan, was footed by the caja real of Veracruz.[101]

12   Survey of the Antigua River in August 1778 and Inspection of a Pine
      Forest in Mecacalco, Santa María Tlapacoya, and Nautla Rivers in
      September 1778

José Vicente Martínez and Cayetano Rangel, who left to survey the region of Antigua, reported to Miguel del Corral and the administrator Pedro Antonio de Cosio. Their report made a detailed geographical, geological, and physical description of the river. Their expedition set off from the so-called Paso de Gallina, following an ancient driveway. They noted the presence of an islet covered in trees which, in their opinion, should be cleared and the timber used to build docks and barges. Del Corral's assistants were so meticulous that they even costed the different stages of the operation. For instance, the clearing of fig trees and willows would cost 40 pesos; the clearing of a brushwood area, "approximately half a mile long and 15 varas wide, the extension needed to open the river pass", was estimated at 100 pesos.[102] In addition to this, the surveyors described every stone, rock wall, and geological outcrop, including their position in the riverbed, size, and estimated weight. They made recommendations as to how these should be worked to bring the river to the necessary depth. The report also includes information on tributaries, currents, and the whirlwinds that formed in the area of Antigua. This description was full of useful details:

---

99   "entre los que los más proporcionados para extraerse son los que se hallan en los sitios
      llamados Chololoya y Zapotitlán, los primeros que deben salir por el Río de la Antigua, y
      los segundos por el de Santa María Tlapacoya y la Barra de Nautla, la extensión de estos
      pinos, y su abundancia es constante". Ibid., f. 156v.
100   Ibid., f. 161v.
101   Ibid., fs. 179–181.
102   "de largo como de media milla y de ancho como quince varas que será lo necesario el
      desmontar para franquear el río". Ibid., fs. 182v–183.

"in Paso de la Barranca there is a nasty turn, [where] a whirlwind, which the locals call *Tragadero*, forms".[103] The river course was surveyed between 12 and 19 August, and the conclusions were not as positive as the viceregal authorities and the *Marina Real* had expected. José Vicente Martínez and Cayetano Rangel argued that transporting wood in Antigua would be impossible because of the costs involved in cleaning the river and regulating the flow.

After the Antigua River, the surveyors took a break, until they began the survey of Mecacalco along the course of the Nautla River on 5 September 1778. On the first day, they visited the slopes of Monte Tadquapan, which was a league and a half distant from the lumberjacks' ranches. They identified and noted several species suitable for masting, between 45 and 28 cubits long (see Table 44).

TABLE 44   Trees identified as useful for masting in Monte Tadquapan

Number of pieces	Length in cubits	Hands in diameter
1	37½	12
1	31	16
1	36	12
1	45	13
1	43½	14
1	44½	14
1	45	13
1	42	10
1	42	18
1	43½	15
1	40½	15
1	30	12
1	43½	15
1	28	13
1	37½	16
1	37½	15
1	30	16
1	31½	14

SOURCE: AGNM, INDUSTRIA Y COMERCIO, VOL. 10, EXP. 10, F. 210

---

103   "en el Paso de la Barranca es una vuelta muy obligada [donde] hay un gran remolino que los pescadores llaman *Tragadero*". Ibid., f. 186.

Although the wood met good standards, the path down which they would have to be dragged did not, as it was traversed by a number of ravines that made transit hazardous. Afterwards, the surveyors visited another nearby elevation, where they noted "this is not a good place for pine, because the trees are dispersed, and most of them have been tested with axes at the base by the locals to see if they can be used for *tajamanil*; as a result, most of them are hollow".[104] This was not a new problem; previous surveys indicated that in other parts of the Sierra Madre Oriental there were clear signs of careless woodcutting by the locals, which ruined many trees that would have been suitable for naval construction. José Vicente Martínez and Cayetano Rangel also pointed out that in other areas there were traces of fire, used by the same locals to clear the undergrowth, which also had a negative effect on the trees. On a hill near Tadquapan, the surveyors felled a young, untouched tree to examine its properties. The wood proved to be unsatisfactory because the core of the trunk was full of fungi and hollow.[105]

During the three following days, the explorers visited two hills between the Sosoguilco River and the ranches of Mecacalco. There they found traces of the survey undertaken by Joseph Benavides and Colonel Miguel del Corral four years earlier. They noted that pines were scarce in that area and headed towards the northeast. In the place of Pila, they found a hill covered in good trees, between 29 and 34 cubits tall (see Table 45). Again, the problem was to take them to the river, because the only available path was steep and rocky.[106]

Therefore, the expedition to Mecacalco did not yield the expected results. The appropriate trees were inaccessible, and those that were well located were too few or had been damaged by the locals, making them unsuitable for shipbuilding. However, Martínez and Rangel did not give up and went on to survey the Santa María Tlapacoya River and the bar of Nautla, setting off from a place known as Tallo on 9 September 1778. Like in Antigua, the surveyors filled their report with interesting geological data about the tributaries and described rock types and sizes in detail. This survey lasted until 21 September, and in addition to noting obstacles (islets, waterfalls, trees, whirlwinds) Martínez and Rangel measured the width and depth of the rivers. Like in Antigua, the conclusions were not positive, owing to the abundance of natural obstacles; similarly, they

---

104   "este paraje no es terreno de Maderas de Pino, pues los que se hallan están muy salteados, y los más tuertos, y todos a pie hachados de los Naturales para reconocer si la hebra es apropósito para rajar para Tajamanil y los más de estos se hallan por dichos hachazos huecos". Ibid., f. 210v.

105   AGNM, Industria y Comercio, vol. 10, exps. 1 y 5.

106   Ibid., exp. 10, f. 212.

TABLE 45   Trees identified as useful for masting in La Pila

Number of pieces	Length in cubits	Hands in diameter
4	30	14
4	30	12
1	29	14
2	30	13
1	31½	16
1	37½	16
1	29	15
1	31½	13
1	33	14
1	31½	12
1	33	12
1	34	16

SOURCE: AGNM, INDUSTRIA Y COMERCIO, VOL. 10, EXP. 10, F. 212

argued that the bar of Nautla did not meet the conditions to house transport vessels.[107] On 28 October 1778, Colonel Engineer Miguel del Corral submitted the final report for the viceroy, pointing out that he had spent a long time talking with the expeditioners and reading their diaries, reaching the conclusion that, although the quality of the wood was better than that in Mecacalco, the river obstacles recommended that further surveys targeted the river and bar of Coatzacoalcos. He also mentioned that the masting requested by the navy commander in Havana could be cut in Tacojalpa, 8 leagues from the river mouth.[108]

13   Projects to Source Wood in Other Regions in New Spain, Such as the Usumacinta River and Laguna de Términos

In 1776, when Miguel del Corral and Joaquín de Aranda were seeking hardwoods suitable for masting in the hills of Perote, Coatzacoalcos, and Oaxaca, the governor of the Presidio del Carmen, Pedro Dufau Maldonado, launched the idea of exploiting the forests near Laguna de Términos (see Figure 11) for naval construction, and of establishing a shipyard at the expense of the

---

107   Ibid., fs. 212v–227.
108   Ibid., fs. 236–237v.

Real Hacienda in Tris Island: "to build frigates and other ships from 24 to 30 cannons".[109] After he submitted a formal proposal to Viceroy Antonio María de Bucareli y Ursúa in the same year, the viceroy endorsed the plan enthusiastically and made the fleet commander in Havana, Juan Bautista Bonet, send officers to explore the area around Laguna de Términos. The survey was undertaken by shipyard foreman Luis Fernández and Frigate Sub-Lieutenant Miguel Sapiain, who only approved the extraction of hardwoods and cedar to send to Havana, while ruling out the possibility of building a shipyard in the Presidio del Carmen.[110] It appears that trees were felled during the three following years, as reflected in a letter by governor Pedro Dufau Maldonado to the viceroy, dated 21 January 1779; the letter indicates that, between 1776 and 1779, Havana had requested 1,000 curved pieces, and that, in 1777, 400 pieces were dressed under the supervision of a private contractor, master shipbuilder Joseph Nicolás Sánchez. The pieces requested included futtocks; square yards; crotches; beams; cedar or mahogany ribbands; a right cedar waterway; a jabí keel; a cutwater; and a mahogany pump.[111] Most of these parts were stored in the Presidio del Carmen and, owing to the scarcity of available shipping, dispatched piecemeal in both private and royal ships when the occasion arose. In a document dated March 1779, Pedro Dufau Maldonado reported that 245 dressed parts were still waiting in the presidio, despite the fact that four loads had been dispatched to Havana between 29 July and 22 December 1778.

The shortage of shipping was a chronic problem during this period, one to which the Crown failed time and time again to find a solution. For this reason, many parts, masts, and trunks were ruined, being left at the mercy of the elements for too long.[112] It is also of note that the initiative of the governor of the region around Laguna de Términos and the Usumacinta River led to the marking and the felling of 15,000 trees, although this official clarified that the works undertaken theretofore and those that were projected for the future, "be it bringing timber to the river for floating, be it dragging down long paths, cost the king very dear".[113]

---

109   "para construir fragatas y otros buques desde 24 hasta 30 cañones de fuerza". AGNM, Correspondencia de Diversas Autoridades, vol. 29, exp. 59, f. 195.

110   Ibid., fs. 204–205v, 222–223.

111   The order was increased by another 181 pieces, coming to a total of 581 dressed parts. AGNM, Real Caja, vol. 15, fs. 25–28v.

112   Rafal Reichert, "El transporte de maderas para los departamentos navales españoles en la segunda mitad del siglo XVIII", *Studia Historica* 43 no. 1 (2021), 47–70.

113   "sea caer maderas al río en que se embalsan, sea el tiro demasiadamente largo, resultaban costosos y sumamente gravosos al Rey". AGNM, Industria Artística y Manufacturera, vol. 2, exp. 3, f. 37.

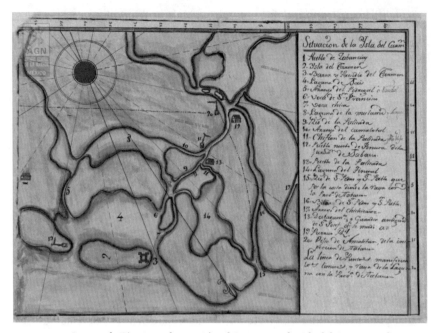

FIGURE 12   Laguna de Términos, the *presidio* of Carmen on the Isla del Carmen and main
             villages and rivers (1792). Made by Juan Manuel de Bonilla
             SOURCE: ARCHIVO GENERAL DE LA NACIÓN DE MÉXICO, MAPAS, PLANOS E
             ILUSTRACIONES (280), MAPILU/210100/3504 ISLA DEL CARMEN (3256)

When the governor of Presidio del Carmen, Pedro Dufau Maldonado, realised the obstacles facing the sourcing of hardwood for naval construction in his
jurisdiction, he presented the viceroy with the proposal of using forest resources
from the Usumacinta River, which he had surveyed previously, cutting down
some trees; his informants had reported the presence of abundant quality hardwoods in accessible areas. However, the vast woodlands near the Usumacinta
fell under the jurisdiction of the governor of Tabasco. Using the argument that
this could keep a steady supply of wood for the navy in Havana, Pedro Dufau
Maldonado requested the authority to exploit that area of Tabasco in 1779.[114]

In his account, Dufau Maldonado reported difficulties with his labour force
"as the hands needed are many, it is not rare for several to run away at a time",[115]
claiming that they sought, and found, shelter with the villagers of Tabasco,
helped by the slack vigilance imposed by the royal officials in the region. The

114   AGNM, Industria Artística y Manufacturera, vol. 2, exp. 3, f. 38.
115   "siendo grande el número de los individuos de que se forma el trabajo, acontece frecuen-
       temente que desertasen varios mozos".

governor of Presidio del Carmen argued that, with his previous experience felling trees in Laguna de Términos, he was especially well qualified to supervise the works.[116]

This appears to have been a widespread problem, for similar complaints can be found elsewhere, for instance during the royal works in Chimalapas in 1767–1768, when the local workmen often fled when the time came to drag the timber to the banks of the Coatzacoalcos River because it was heavy work. Many hands escaped after collecting their wages to tend to their fields and crops.[117]

It is also interesting to note that Pedro Dufau Maldonado accused the locals of felling "trees clandestinely for their own use, and the governor of Tabasco cannot stop it, because the hills are inaccessible, hard to navigate, deserted, and full of streams and debris".[118] His petition closed with the argument that, if no permission was granted to move to the region around the Usumacinta River, he should at least be allowed to supervise tree felling in the region.[119]

Viceroy Antonio María de Bucareli y Ursúa's reaction was to address the governor of Tabasco in late March 1779, whom he compelled to "look after the trees and ensure that none were felled near the Usumacinta, punishing those who do".[120] He also requested support for Pedro Dufau Maldonado's work in Laguna de Términos. The governor was also asked to report on the "abundance and quality of timber on the Usumacinta's banks [...] so that full information becomes available about woods suitable for naval construction, and how difficult or easy it should be to bring it to Havana".[121] In any case, the viceroy did not extend the jurisdiction of Presidio del Carmen over the region of the Usumacinta, simply asking the governor of Tabasco to help "with the necessary supplies, and with all the other things requested".[122]

---

116   AGNM, Industria Artística y Manufacturera, vol. 2, exp. 3, fs. 39–40.

117   AGNM, Industria y Comercio, vol. 10 exp. 1, f. 17.

118   "los cortes furtivos de maderas para intereses propios de algunos sujetos particulares, y es dificultosísimo que pueda impedirlo el gobernador de Tabasco, pues en la actualidad acontece que sin su permiso se están cortando, y sacando maderas a causa de ser los montes inaccesibles intrincados, desiertos, llenos de raudales y desechos".

119   AGNM, Industria Artística y Manufacturera, vol. 2, exp. 3, fs. 40–41.

120   "cuide, cele y vele, no se ejecuten cortes de maderas en los ríos de Usumacinta, castigando a los que lo ejecutaren".

121   "la abundancia, y calidad de maderas de las riberas de aquellos ríos de Usumacinta [...] para tomar, una radical y plena instrucción de las maderas de aquel país, si son abundantes de solides, y calidad, propia para fábrica de navíos y de facilidad, o dificultad que pueda haber de no extracción de aquel país, y conducción para la Habana, y porque conducto pueda practicarse informador con toda justificación sobre este asunto".

122   "auxilie con víveres necesarios, y en lo de más que le va prevenido". AGNM, Industria Artística y Manufacturera, vol. 2, exp. 3, f. 42.

Summing up, the 1770s witnessed the implementation of several projects to survey the wood available in the regions of the Seno Mexicano, between Veracruz and Campeche, in order to boost the wood supply for the naval departments in Havana and the Iberian Peninsula, as well as the warehouses in Veracruz and Campeche. Not all of these projects had a positive outcome for the *Armada Real* because in many instances the inaccessibility of woodland areas, natural obstacles, problems with labour and draught animals, and the difficulties of storing ship parts on riverbanks and the seashore made them too costly. However, these ideas, also promoted by local governors, as illustrated by the governor of the Presidio del Carmen, Pedro Dufau Maldonado, greatly contributed to increasing the Spaniards' topographic and geographical knowledge of these regions, which had theretofore been explored only lightly. In conclusion, the expeditions undertaken by Spanish officials and their agents in Nautla, La Antigua, Perote, Mecacalco, Alvarado, Chimalapas, Coatzacoalcos, Usumacinta, and Laguna de Términos led to a much better understanding of the wood resources available in the provinces of Veracruz, Oaxaca, Tabasco, and Campeche (see Map 9 and Figures 13–15).

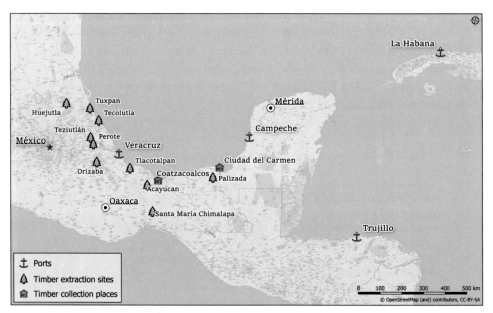

MAP 9   Forests, storage sites, and ports where timber harvested in the provinces of Oaxaca, Veracruz, and Yucatan was shipped.
SOURCE: LOOK4GIS-LUKASZ BRYLAK, KAROLINA JUSZCZYK AND DANIEL PRUSACZYK BASED ON QGIS SOFTWARE

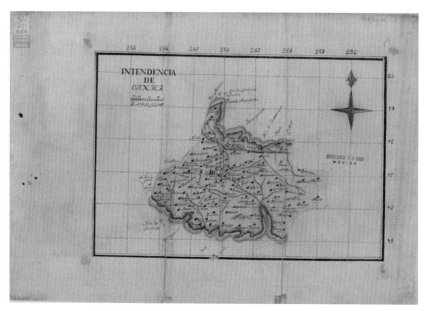

FIGURE 13　The map shows the territorial draft of Intendencia of Oaxaca (1774)
SOURCE: ARCHIVO GENERAL DE LA NACIÓN DE MÉXICO, MAPAS, PLANOS E
ILUSTRACIONES (280), MAPILU/210100/88 INTENDENCIA DE OAXACA (84)

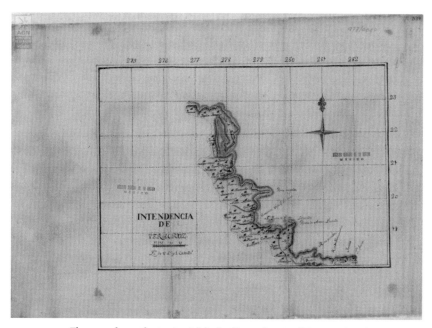

FIGURE 14　The map shows the territorial draft of Intendencia of Veracruz (1774)
SOURCE: ARCHIVO GENERAL DE LA NACIÓN DE MÉXICO, MAPAS, PLANOS E
ILUSTRACIONES (280), MAPILU/210100/89 INTENDENCIA DE VERACRUZ (85)

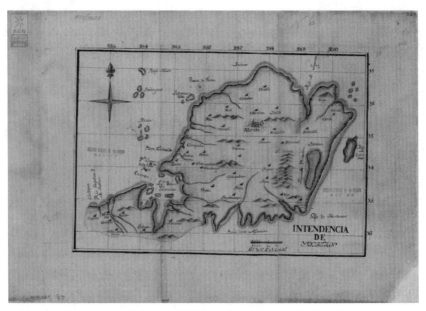

FIGURE 15   The map shows the territorial draft of Intendencia of Yucatan (1774)
             SOURCE: ARCHIVO GENERAL DE LA NACIÓN DE MÉXICO, MAPAS, PLANOS E
             ILUSTRACIONES (280), MAPILU/210100/87 INTENDENCIA DE YUCATÁN (83)

14   Wood Species in New Spain, Based on Subdelegate Pedro Cabezas's Survey of Huejutla

The survey of the New Spanish territory in search of wood resources continued to the first decade of the 19th century for the king's naval needs, for local uses, and in support of the mining industry. This resulted in several reports issued by royal officials in Morelia, Guanajuato, San Luis Potosí, Zacatecas, Puebla, and Oaxaca in the 1790s.[123] Especially outstanding is a report that inventoried all the tree species available in the region of Huejutla, written in March 1790 by the region's subdelegate Pedro Cabezas, appointed by the general intendent of the province, Bernardo Bonavia, to find samples of valuable trees. Cabezas explained that he had taken small samples of wood so that their density, colour, veins, and other physical features could be examined.[124] He described the following species:

---

123   AGNM, Industria y Comercio, vol. 31, exps. 13, 19, 20, 21, and 24.
124   Ibid., exp. 10, fs. 239–243v.

1. *Smoketree*: "this tree is approximately 20 varas tall, and the trunk is approximately 1 vara in diameter. The top is thorny, and the leaves, and even more the flower, are tiny. It produces curly pods, like a snail, and the locals commonly use this fruit. The bark of the trunk is useful to tan leather and even to give cloth a green colour with red veins."[125]

2. *Chijol*: "this tree is approximately 20 varas tall, and the trunk is approximately 1 vara and a half in diameter. It is used for local construction, as the buildings in the region are generally made of timber. The chijol gives them beams and other parts, which have the advantage of being impervious to moths and damp. Moreover, the parts that get buried in damp soil petrify (as they say around here). The wood is harder than holm oak's, and although I have not weighed it, it can be seen that it is as dense, or denser."[126]

3. *Volantín*: "this tree is approximately 30 varas tall and 1 vara in diameter, and it is used for roofs, for it is very hard. It is called volantín because, as the trunk is tall and straight, the Indians use it for the game of volantín: they stick a trunk on the ground, after removing all branches, and on the top they place a cross, from whose arms four ropes hang; one Indian, at the end of each rope, run around the trunk and in this way climb up, and it is something to behold that, while the four dance below, there is another dancing on top of the trunk in a little table barely big enough to set a foot: this is a barbarity that I have never allowed in my jurisdiction."[127]

---

125   "es un árbol como de 20 varas de altura y su tronco tiene por lo regular como una vara de diámetro. Las ramas de la copa todos son espinosas. La hoja es menuda y mucho más la flor. Produce unas vainas enroscadas como caracol y esta es fruta corriente de que usan por aquí. La cascara del tronco es muy útil para curtir pieles y aun para dar a las gamuzas un color verde con vetas rojas".

126   "es un árbol como de 20 varas de alto y una y media de diámetro. Es muy útil para los edificios de estos países, que por lo regular son de madera. Pues del Chijol sacan horcones, vigas, y demás piezas las que tienen la ventaja de no entrarles jamás polilla, ni podrirse con la humedad. Pues antes se petrifica (según asientan aquí todos de voz común) la parte que se enclava en terreno húmedo. Es esta una madera más dura que la encina y aunque no he tenido la curiosidad de calcular su peso pero hecho de ver que es de igual o mayor gravedad especifica que la de la encina".

127   "es árbol como de 30 varas de altura y una de diámetro y se aprovecha para techumbres de edificios por ser madera muy maciza. Le dicen volantín porque con motivo de ser su tronco muy recto y alto usan de él los Indios para el juego llamado Volantín que consiste en poner fijo o clavado uno de estos Palos (quitado toda la ramificación) y puesta en la punta una cruceta que voltea fácilmente, penden de sus cuatro brazos, cuatro cuerdas en cuyos extremos se sientan otros tantos Indios y dan vueltas en contorno, con cuyo movimiento van adquiriendo una grande elevación siendo lo más digno de admirar que ínterin dan vueltas los cuatro Indios de las cuerdas, esta otro bailando sobre la punta del madero en una especie de mesilla en cuyo plano apenas cabe la planta de un pie: barbaridad que nunca he permitido yo en este partido".

4. *Higuerón*: "this tree is between 20 and 25 varas tall and 1 vara in diameter, and the wood is not used, only to extract a sort of milk that it exudes, when the bark is cut, to treat hernias. The tree yields a fruit similar to the fig, but smaller."[128]

5. *Hule*: "this tree is up to 25 varas tall and over 1 vara in diameter. The wood is also not used, only the milk that it exudes (also called hule), which is used to tar shoes, capes, and other clothing for the rainy season. It is said that the milk is also used in poultices in disjointed bones, so they do not go out of place again".[129]

6. *Tzocohuite*: "this tree is approximately 15 varas tall and 1 vara in diameter. The wood is good for nothing, except to make charcoal for ironsmiths."[130]

7. *Quiebrahachas*: "this tree is between 30 and 35 varas tall and one and a half in diameter. The wood is dark, as shiny as mahogany, but locals do not use it because it is too hard, hence the name quiebrahachas, and dressing it is difficult; it as dense or denser than holm oak."[131]

8. *Jonote*: "this tree is approximately 8 varas tall and three-quarters in diameter. The wood is extremely light, but this makes it suitable for roofbeams, being impervious to moths; it must be kept away from the damp, or it rots."[132]

9. *Sangre de dragón*: "approximately 20 tall and three-quarters in diameter. When cut, the trunk gives away a blood-red liquid (from which it may have taken the name), sold by apothecaries for various maladies."[133]

---

128 "es un árbol de 20 a 25 varas de alto y una de diámetro y aunque para nada aprovechan su madera, pero utilizan una leche que vierte (picando el tronco verde) para curar a los relajados o quebrados. Da este árbol una fruta de la figura de la breva, pero más pequeña y su sabor es parecido al del higo".

129 "es un árbol hasta de 25 varas de alto y más de vara de diámetro. Tampoco se aprovechan de su madera sino de una leche que vierte (llamada también hule) con la que embetunan zapatos, capingones y otras ropas a efecto de defenderse del agua en tiempo de lluvias. Se dice que también sirve esta leche para bilmas que se aplican a los que han tenido huesos dislocados para que no vuelvan a dislocarse".

130 "árbol como de 15 varas de alto y una de diámetro cuya madera no sirve más que para carbón de herreros".

131 "es de 30 a 35 varas de alto y su tronco como de vara y media de diámetro. Es madera negra tan lucida como el ébano pero no la aprovechan para piezas porque es tan sumamente dura que por eso le llaman quiebrahachas y cuesta mucha dificultad el labrarla fuera de que también es tan pesada o más que la encina".

132 "como de ocho varas de alto y tres cuartas de diámetro. Su madera es sumamente leve cuasi tanto como el corcho y sin embargo esto sirve para vigas de las casas porque no le entra polilla y es de mucha duración en no teniendo humedad porque con el agua luego se pudre".

133 "como de 20 varas de alto y tres cuartas de diámetro. Su tronco en dándole un tajo vierte un licor rojo, color de sangre (que acaso de aquí tomó el nombre este árbol) que se vende en las Boticas para que la aplican los Médicos para varias enfermedades".

10. *Chaca* or *palo mulato*: "approximately 20 varas tall and half in diameter. It is deemed to be very fresh, and the boiled leaves are used as a remedy against fever."[134]

11. *Brasil*: "from 15 to 20 varas tall and over half in diameter. The wood is very hard, being used for roofs and by dyers to get a purple colour."[135]

12. *Tepeguage* or *vochiscoahuite*: "approximately ten varas tall and two in diameter. The wood is very hard, and locally it is used only in sugar mills."[136]

13. *Chicozapote*: "approximately 25 varas tall and one and a half in diameter, the fruit is well known. When the trunk is cut, it gives out a milk vulgarly known as chicle. The wood is used in building, and the least thick branches on staffs, which are highly appreciated owing to their red colour and softness; they are also used for bottoms."[137]

14. *Mulberry tree*: "approximately 20 varas tall and half in diameter. The wood is soft and a fine yellow colour, used to dye capes. The fruit is called the mulberry, but it is different from those picked in cold regions because it looks similar, but the flavour is different, as well as being larger and green in colour."[138]

15. *Copal*: "approximately 12 varas tall and nearly 1 in diameter. The wood is only good for firewood, but it exudes a gum or aromatic resin that is used in temples instead of frankincense."[139]

16. *Cedro*: "up to 25 varas tall and up to 2 in diameter. The wood is very useful, not only for beams and roofs, but also for chairs, chests, wardrobes, and other common pieces of furniture, and especially for frames, statues, and altarpieces. There is a great abundance of these trees in the region, so

---

134 "como de 20 varas de alto y media de diámetro. Se dice ser muy fresco y cocidas las hojas en agua común se usa esta bebida para destruir las fiebres".

135 "de 15 a 20 varas de alto y más de vara de diámetro. Es madera muy dura que sirve para techumbres y la usan los tintoreros para dar color morado".

136 "como de diez varas de alto y dos tercios de diámetro es madera muy dura que por aquí solo se aprovecha para los trapiches o molinos de caña".

137 "como de 25 varas de alto y una y media de diámetro cuya fruta es muy conocida. Hendido su tronco vierte una leche que vulgarmente llaman chicle. Su madera se aprovecha para edificios, y las varas menos gruesas sirven aquí para báculos que son muy lucidos por su tersidad y color rojo y pueden aplicarse también para tacos de trucos".

138 "como de 20 varas de alto y media de diámetro. Su madera es muy tersa de un color amarillo muy fino que sirve para teñir Gamuzas. Su fruta se llama mora pero es distinta de la de tierras frías pues aunque se parece en la figura pero es de diverso sabor de mayor tamaño y de color verdoso".

139 "como de 12 varas de alto y cerca de una de diámetro. Su madera no sirve más que para leña, pero vierte una goma o resina odorífera que molida en polvo sirve en los Templos en lugar de Incienso".

much that they are also used for benches, steps, doors, windows, balconies, basins, and such, but it is not as fine as the cedars found in Havana."[140]

17. *Chiote*: "approximately 16 varas tall and one and one-quarter in diameter. The wood is good for nothing, but it gives out a fruit called Guajilote, like a banana from Guinea, but with sharper ends. The Indians eat it regularly and they say it is useful for different maladies and especially to facilitate menstruation."[141]

18. *Palo de rosa*: "approximately 18 varas tall and three-quarters in diameter. The wood is hard and elastic, so it is used for beds and roofbeams in houses. Around March, it loses most leaves but sprouts a multitude of flowers (white in colour, with ends like a rose) so that every tree looks like a bunch of flowers. The flowers make figures, like buttons and leaves. Five flowers spring from each branch, the two below facing the opposite side as the three above."[142]

19. *Orejón*: "approximately 10 varas tall and three-quarters in diameter. The wood is good for nothing, but the bark is used to line ramrods. The leaves are tiny."[143]

This report had little effect, as Huejutla's wood resources remained largely untapped in the late 18th century, but is a valuable illustration of the relationship between humans and their ecological context. Pedro Cabezas made a thorough description of the types of wood available in the province, including the uses to which they were put: medicinal – ritual; in the textile and footwear

---

140   "como de 25 varas de alto y hasta de dos de diámetro. Es madera muy útil no solo para Viguerías y techumbres, sino para Sillas, Baúles, Roperos, y demás piezas del uso común y especialmente para toda labor de talla como Óvalos, Estatuas, Retablos. Hay de estos árboles mucha abundancia por aquí tanto que no se usan de otra cosa todos los muebles de Bancas, Escaños, Puertas, Ventanas, Balcones, Tinas, Bateas, y demás, pero no es este cedro tan fino como el de La Habana".

141   "como de 16 varas de alto y una y cuarta de diámetro. De nada sirve su madera pero da una fruta que llaman Guajilote de figura de plátano guineo, aunque con los extremos más agudos, la que comen regularmente los Indios, y se dice ser muy útil para varias enfermedades especialmente para facilitar la menstruación de las Mujeres".

142   "como de 18 varas de alto y tres cuartas de diámetro. Su madera es muy elástica y dura por lo que se aplica para catres y también suelen hacer de ella las Vigas de las Casas. Por el mes de marzo se le cae la mayor parte de las hojas pero produce una multitud de flores (color blanco con los extremos aportillados como la Rosa) de suerte que cada árbol parece un Ramillete. Va dibujado un ramo con la figura de las flores, botones y hojas. De estas nacen cinco en cada ramito, las dos inferiores con inclinación opuesta a la de las tres superiores".

143   "como de diez varas de alto y tres cuartas de diámetro. Su madera de nada sirve pero la corteza del tronco se aprovecha para curtir Baquetas. La hoja es muy menuda". AGNM, Industria y Comercio, vol. 31, exp. 10, fs. 239–243v.

industry; and in the construction of buildings and furniture. In some passages, Cabezas mentioned the local communities, on whose knowledge of these trees the Spaniards plainly relied, making use of it for their own purposes. This and similar reports written in the 1790s betray a change of perception in wood use. The trees were no longer only seen as a possible source of naval timber; other uses were also contemplated. In addition, reports written in mining regions began warning about deforestation, the result of overexploitation and cattle ranching.[144] Finally, concerning the matter of naval construction, Cabezas's opinion that the cedars from Huejutla were not "as fine as the cedars found in Havana" is of note.[145] This reference suggests that the subdelegate had previous experience with wood in other regions, even distant ones, like Cuba, where cedars were chiefly used by the naval industry.

---

144   Ibid., exp. 13, fs. 258–265v.
145   "tan fino como el de La Habana". Ibid., exp. 10, fs. 241v–242.

CHAPTER 4

Timber *Asientos* for the Spanish Royal Navy in New Spain

The final chapter deals with logging contracts in the Gulf of Mexico and the Caribbean Sea region for the supply of the Spanish naval departments in the second half of the 18th century. These operations, carried out in several locations in the Greater Caribbean, were the direct consequence of the policy pursued by the Crown to systematically upgrade its *Marina Real* during this period. The forest surveys mentioned in Chapter 3 were instrumental for these contracts, which included the felling, processing, and supply of timber for shipbuilding.

The *asientos* granted by the colonial viceroys and governors secured the supply of cedar and mahogany for the shipyards. However, the analysis of the contracts shows that the main beneficiaries of these *asientos* were the local elites, who reaped handsome profits and expanded their political and economic influence in their provinces.

1 Early Timber *Asientos* in the Viceroyalty of New Spain in the Second Half of the 18th Century

The survey projects sponsored by viceroys Croix and Bucareli had little impact on the Secretary of the Navy, as the only large-scale felling operations for masting were those in the region in Chimalapas, where pine was sourced for the naval departments in Havana and Spain. These operations were supervised by royal officials and directly funded by the *cajas reales* in New Spain. The viceroyalty's limited contribution to supply was due to the small number of *asientos* for the felling and processing of timber awarded to that date, and to the small demand for naval timber in New Spain. Despite this, some contracts were negotiated in the second half of the 18th century; these aimed to supply fleets, squadrons, shipyards, and the navy departments with prepared spare parts, bottoms, and full vessels.

The first known such *asiento* in New Spain was awarded to Captain Andrés Berdeja, for the supply of dressed parts to the Armada de Barlovento in 1742. The contract was still valid by 1748, when, as previously noted, Diego Fentanez, the owner of the haciendas of Solcuautla and Cuatotolapán, filed a lawsuit against the *asentista* for felling some cedars without permission.[1]

However, it was not until the 1760s that the first large-scale operations to supply the *Marina Real* were undertaken. The first *asiento* of the decade was awarded to Brigadier General Antonio Basilio Berdeja, representing his father Captain Don Andrés Berdeja, in 1766. This contract heeded an order issued in 1764 by Viceroy Marquis of Cruillas in response to the the king's fleets in Veracruz's scarcity of wood. It was at the time not rare to sell beams and pieces of masting to private customers "to build houses".[2] In December 1765, the viceroy decided to award a long-term contract to experienced *asentistas*, such as Andrés Berdeja and his son Antonio. Initially, the contract had a duration of five years, beginning in February 1766, and contemplated the supply of prepared mahogany and cedar parts to the royal warehouses of Veracruz to make, stocks, tillers, boards of different sizes, and curves. The trees were felled in the hills of Tlacotalpan.[3] The first cargo was delivered in June 1766, and consisted of the following:

- a rudder, 14 *varas* long, 36 inches across, 18 inches thick, and 25 inches square;
- a tiller of zapote, 10.5 *varas* long and 13 inches square;
- a stock 7½ *varas* long and 20 inches across;
- a stock 7 *varas* long and 21 inches across;
- a stock 7¾ *varas* long and 20 inches across;
- a stock 5½ *varas* long and 22 inches across;
- a curve, larger than usual;
- 16 knees, regular size;
- three mid-sized curves;
- 25 large planks;
- 400 regular boards;
- 52 regular side boards;
- 369 small boards;
- 98 costons;
- 13 beams.[4]

---

1 Rafal Reichert, "¿Cómo España trató de recuperar su poderío naval? Un acercamiento a las estrategias de la marina real sobre los suministros de materias primas forestales provenientes del báltico y Nueva España (1754–1795)", *Espacio Tiempo y Forma. Serie IV, Historia Moderna* 32 (2019), 82.

2 "obras de reedificación de casas". AGNM, Correspondencia de Diversas Autoridades, vol. 8, exp. 105, fs. 391–392.

3 Rafal Reichert, "Recursos forestales, proyectos de extracción y asientos de maderas en la Nueva España durante el siglo XVIII", *Obradoiro de Historia Moderna* 28 (2019), 71.

4 AGNM, Industria y Comercio, vol. 10, exp. 2, f. 60.

This delivery had a cost of 1,882 pesos and 3 reales. More deliveries of dressed pieces and boards followed, and all the parts delivered in 1770 fetched a cost of 9,173 pesos, paid by the *caja real* of Veracruz. This brought the first major *asiento* awarded by the Crown to members of the Veracruz elite, in the second half of the 18th century. It is important to note that the area exploited by the Brigadier General was visited by Miguel del Corral during his survey of Veracruz and Oaxaca in 1777. In his description, del Corral pointed out that he had found traces of the *asentista*'s woodmen's work.[5]

The document does not specify the sort of wood used, but other documents suggest that the stocks and the beams were made of hard mahogany and the boards of cedar. The use of zapote for a tiller is of great interest because this wood was rarely used in shipbuilding, although it was frequently employed in construction; Joseph Jiménez was hired to supply fruit trees and cedar wood for the renovation of the Castle of San Juan de Ulúa, in Veracruz, in 1775. When he began felling trees in Tuxpan, Tecolutla, and Papantla, Jiménez found out that the shipbuilder Juan Felipe Michelena was felling zapotes without authorisation. This illegal operation was stopped by the *alcalde mayor*, Juan Joseph Enciso, to whom the locals had complained. The official explained that he had intervened because, apart from the fact that the felling was unauthorised, "the Indians make use of the zapote, not only eating its admirable fruit but for that resin that they call *chicle*, the crop of which sustains them and helps them to pay their taxes".[6] This conflict between Joseph Jiménez and Juan Joseph Enciso, like that which pitched Diego Fentanez against Andrés Berdeja in 1748, illustrates the problems that ensued when *asentistas* abused their rights and sourced wood outside the designated *realengo* forests, clashing with local *hacendados* and indigenous communities.

2   An Attempted *Asiento* for Masting in Favour of Domingo Ramón Balcarsel in 1768

It has been noted that in addition to the operations supervised by royal officers in Chimalapas, a private *asiento* to fell pines in Veracruz was also awarded in the 1760s. From mid-1767 onwards, the royal officials of Veracruz and the navy

---

5   Alfred H. Siemens and Lutz Brinckmann, "El sur de Veracruz a finales del siglo XVIII", 311.

6   [los Zapotes hacen el bien de los Indios, como que de ellos utilizan no solo en su estimable fruto con que se alimentan, sino en él de sus resinas y jugo de que sacan el que llaman chicle, en cuya cosecha fundan mucha parte de sus alivios y la paga de Tributo]. AGNM, Industria y Comercio, vol. 10, exp. 6, fs. 112–124v.

were trying to find a way to provide the local warehouse and the navy departments with masting, which was in short supply. Early in 1768, Don Domingo
Ramón Balcarsel offered his services to make up for these shortcomings.[7] The
contractor committed to extracting from Perote the full masting for three
80-gun ships for 12,000 pesos de a ocho, three 70-gun ships for 11,000 pesos de
a ocho, four 60-gun ships for 10,000 pesos de a ocho, and six 30-gun frigates
for 5,500 pesos de a ocho. In order to make his proposal more attractive for the
Real Hacienda, the asentista offered an 8% discount over the total price. The
asiento was approved by Viceroy Marquis of Croix and, on 25 July 1768, by the
king. The royal order was sent to New Spain with a list made in the shipyard of
La Carraca by Don Ciprian Autran, who had also assessed the proposal, with
the number and dimensions of the required masts.[8]

It is known that Don Domingo Ramón Balcarsel began felling on the banks
of the La Lana River. However, the scarcity of good trees for masting—there
were only ordinary pines, known by the natives as ocotes—made him relocate
to a hill that was a league distant from Zapotitlán, where he felled approximately 250 pieces. He then faced problems moving the timber to and down the
river—something about which Lorenzo Arrinda had warned in his report—
forcing the asentista to ask for more money to finish the job. This was denied,
and the contractor forfeited his contract, returning the masting and the tools
that he had been given when the contract was signed. Afterwards, he filed a
lawsuit to get the money spent on wages refunded. Interestingly, engineer
Miguel del Corral had visited these areas in 1777, reporting that the pines there
were too heavy and knotty, which made them unsuitable for masting.[9]

Finally, it is worth pointing out that the 1770s did not witness further significant initiatives to promote the extraction of wood for naval purposes in
New Spain. It is true that Viceroy Bucareli sponsored the survey of woodland in
Veracruz, Oaxaca, Tabasco, and Campeche, but these efforts did not crystallise
in significant felling operations. The asiento signed by the Bordejas in 1766 was
still active by 1773,[10] and the transport of the final 350 masts extracted from
Chimalapas, which were at the time sitting in a tinglado (an organised store
of timber (see Figure 16)) at the mouth of the Coatzacoalcos, was being organised,[11] but no additional contracts for naval supplies were signed during this

7   AGS, SMA Asientos, leg. 624. La propuesta de contrata de arboladuras.
8   AGNM, Industria y Comercio, vol. 10, exp. 5, f. 106v y AGS, SMA Asientos, leg. 624. La
     propuesta de contrata de arboladuras.
9   AGNM, Industria y Comercio, vol. 10, exp. 5, f. 107 and Siemens and Brinckmann, "El sur de
     Veracruz a finales del siglo XVIII", 313–314.
10   Reichert, "Recursos forestales, proyectos de extracción y asientos de maderas", 71.
11   AGS, SMA, Arsenales, leg. 351, oficio núm. 697.

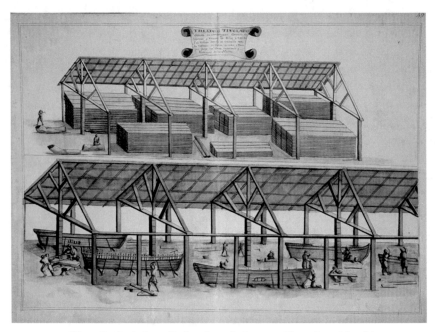

FIGURE 16   View of a *tinglado* or *tillado*, where the boats and launches of the ships were
            built, and painted in order to preserve all the cedar planking for use in the ships
            SOURCE: SPAIN, MINISTRY OF DEFENCE. ARCHIVO HISTÓRICO DE
            LA ARMADA SEDE "ELCANO". *DICCIONARIO DEMOSTRATIVO CON LA
            CONFIGURACIÓN* (SHEET 49)

decade. This was probably due to the new focus on the exploitation of wood
resources in other regions of the Gulf – Caribbean,[12] such as Cuba, Louisiana,
Venezuela, and New Granada. Wood continued being extracted for other purposes, as shown by the *asiento* signed by Don Joseph Jiménez in 1775 to supply
fruit trees and cedar wood for the renovation of the Castle of San Juan de Ulúa
in Veracruz.[13] Similarly, the odd cargo of wood was sent to Spain; for instance,

---

12   The use of this composite term denotes the shared characteristics of the different territo-
      ries in this geohistorical region, such as European colonisation, the use of African slaves,
      and a plantation-based economy. Other names that have been used to refer to the region
      are the Greater Caribbean and Circuncaribe. See John H. Parry and Philip M. Sherlock, *A
      Short History of the West Indies* (London/New York: Macmillan/St. Martin's Press, 1956);
      Sidney Mintz, *Caribbean Transformations* (Chicago: Aldine, 1974); Antonio Gaztam-
      bide-Geigel, "La invención del Caribe en el siglo XX", *Revista Mexicana del Caribe* 1 (1996),
      74–96; Johanna von Grafenstein, *Nueva España en el Circuncaribe, 1779–1808. Revolución,
      competencia imperial y vínculos intercoloniales* (Mexico: UNAM, 1997).
13   AGNM, Industria y Comercio, vol. 10, exp. 6, fs. 112–124v.

three royal ships (*Santa Rita, Visitación,* and *Santa Polonia*) loaded with cedar seized from the private ship *Santa Ana,* entered the harbour of Veracruz in September 1776 with an unregistered load of 309 *tozas* and 161 boards of cedar. The cargo, which came to a total volume of 1,782 square cubits, was sent to El Ferrol on behalf of the *Real Hacienda*.[14]

3   The 1783 Order by the Secretary of the Navy, Antonio Valdés, to Undertake Large-Scale Felling in New Spain

The most important event in the history of wood *asientos* in New Spain during the colonial period took place on 28 July 1783, when the Secretary of the Navy, Don Antonio Valdés (1783–1795), sent Viceroy Don Matías de Gálvez y Gallardo (1783–1784) the order to authorise large-scale cedar-felling operations on the coast of Veracruz, in order to supply the naval departments of Cádiz-La Carraca, El Ferrol, Cartagena, and Havana.[15] The purpose of the order was to boost Spanish naval construction and the wood trade between New Spain and Spain, so that:

the number of our warships is increased and that the Navy is put in a tolerable footing, an essential condition for a Commercial Power whose possessions are scattered over the four parts of the world [...] so that they can come in their defence whenever necessary, preventing them passing to foreign hands, as has happened all too often, [and] Powerful Navies interdicting trade and interrupting our navigation, which is the nerve of Monarchies. [...] Great Britain, whose haughty and proud power leads her to call herself the Mistress of the Seas, tries to usurp the glory that is not hers, abusing her power and natural conditions [...]. The Giant at last fell in the recent War,[16] losing the glorious epithet at the hands of the powerful combined forces of the House of Bourbon [...]. Expecting that England's resentment will finally stew for long enough to push her to seek revenge, [the Bourbon Crowns] are reading themselves so that this does not catch [their fleets] unprepared, seeing that in the Thames and shipyards [the English] do not stop working, repairing their warships and building new ones.[17]

14   Ibid., exp. 9, fs. 137–144.
15   Reichert, "Recursos forestales, proyectos de extracción y asientos de maderas", 74.
16   The American War of Independence, in which Spain participated from 1779.
17   aumento de los bajeles de Guerra y poner nuestras fuerzas marítimas en un pie respetable, indispensables a una Potencia Comerciante cuyos dominios se hallan dispersos

The Secretary of the Navy set out nine points to be taken into account before granting an *asiento*. First, Valdés indicated that "contracts should consider the felling of trees and their transport to the embarkation point; the wood is to be examined so no poor-quality timber is accepted: once embarked and approved, the prices agreed will be paid".[18] Second, the assessment of the wood was to be undertaken by a qualified officer sent from Havana. Interestingly, this point also establishes that both *asentistas* and private persons could deliver wood at the embarkation point "without limitations", which suggests that the *Marina Real* aimed to hoard as much wood as possible, as parts of cedar and other hardwoods were in short supply, without regard for the possibility that mass felling could lead to the deforestation of areas of Veracruz. Valdés also ordered the wood to be paid for at the embarkation point after an expert had measured it. The third point addressed the transportation of masts and dressed pieces to Spain and Havana, in private or navy ships, which was to be funded by the state. When transport fell to private vessels, the viceroy was responsible for negotiating the price, according to the volume of the cargo. The Secretary warned that all efforts should be made for transactions to be beneficial to the king and the *Real Hacienda*. The fourth point dealt with merchants who decided to send wood cargoes to Spain "on their private initiative".[19] The Navy was to retain the right of first refusal, but once this was renounced merchants had the right to freely sell parts, masting, and boards. The fifth point described the woods that had theretofore been used in naval construction, pointing out that the most useful for this purpose were "cedar, mahogany, sabicú, chicharrón, and

---

en las cuatro partes del mundo [...] para acudir con ellas siempre y cuando sea necesario a su defensa, estorbando [que] pasen a dominio extranjero como no pocas veces ha acaecido [...] interrumpiendo nuestro comercio que es el nervio de las Monarquías por las poderosas escuadras enemigas, cuyos buques en los cruceros imposibilitaban nuestra navegación. [...] La Gran Bretaña cuya potencia soberbia y orgullosa se denomina la Señora de los Mares intentando usurparse contra todo derecho esta gloria, abusando de su poder y natural situación [...]. Cayó en fin este Gigante en la Guerra próximamente fenecida quitándole aquel glorioso epíteto que se habían apoderado [de él] las poderosas fuerzas combinadas de la Casa de Borbón [...]. Así previendo que los resentimientos de la Inglaterra algún día se acumularan para satisfacer su enojo, [por eso las coronas borbónicas] intentan aprestarse y prevenirse en tiempo para que no les coja descuidadas [sus escuadras], viendo que en el Támesis y sus astilleros, [los ingleses] no cesan de trabajo, así en carena y composición de sus buques de Guerra como en la construcción de otros. AGNM, Industria y Comercio, vol. 10, exp. 11, fs. 261–262

18   "se formalice contrata par el derribo de árboles, y su arrastre hasta el paraje donde haya de embarcarse, capitulando que será reconocida la madera al tipo de su embarco, para no admitirse la que no sea de calidad: pero embarcada o reconocida por buena, se pagará a los precios que se estipulasen". Ibid., f. 248.

19   "por su cuenta y riesgo".

quiebrahacha; but as the New Spain may contain others that can also be useful, let the responsible officials in Havana decide".[20] The sixth point explained that only the wood supervisor was to be on the king's payroll, while the remaining labour—carpenters, woodmen, and hands—was to be paid by *asentistas* and private agents. Valdés admitted the possibility of paying bonuses to Navy officers to revise the cargoes and give guidance to the people working in felling areas or embarkation points. Points seven and eight declared these officers responsible for the quality of the wood sent. On the other hand, Viceroy Gálvez, as someone who "knows the country and its people",[21] was authorised "so that the king may use the excellent woods of those dominions, to use all means available to his discretion to provide them, giving as many privileges as necessary, as long as they are not detrimental to the *Real Hacienda*".[22] Finally, Secretary Valdés's ninth point emphasised the need for constant communication between the viceroy and the king, to keep the latter abreast of the result of *asientos* and the supply and price of wood. On 17 October 1783, the document was delivered to Viceroy Gálvez, who issued an order concerning wood *asientos* on 7 November.[23]

In his order, the viceroy suggested sourcing the timber near the coast of Veracruz to expedite its delivery to Havana and Spain. To begin with, Gálvez requested a qualified official from Havana to survey the hills in the vicinity of the Alvarado and Coatzacoalcos rivers, as well as other suitable areas. A set of guidelines was to be published giving details of all the parts requested, including quality and size requirements. The viceroy ordered the governor of Veracruz to publish these instructions in all jurisdictions where felling was underway so that all vassals "that wish to enter this business find in V.E. all necessary information; the transactions will be formalised by the royal officials in Veracruz".[24] The viceroy emphasised the importance of thoroughly assessing the quality of the parts on delivery and authorised the purchase of parts to

---

20   "más útiles para aquel fin son cedro, caoba, sabicú, chicharrón o quiebrahacha; pero como puede haber en Nueva España algunas otras clases que también lo sean, se estará para su admisión a lo que diga el facultativo que se destine de La Habana". Ibid., f. 248v.

21   "aquel país y sus naturales".

22   "la ventajosa idea de que el Rey se utilice de las excelentes Maderas que hay en aquellos Dominios facilitando los medios que hallare oportunos a su logro y proporcionado a los que quieran entrar en esta obra cuantos beneficios sean compatibles con el ahorro de la Real Hacienda". Ibid., f. 249.

23   Ibid., fs. 247–249v; AGNM, Reales Ordenes, vol. 3 exp. 37, fs. 51–54v.

24   "que quieran dedicarse a esta negociación, hallarán en V.E. cuanta protección que necesitasen y que para formalizar el ajuste y la contrata ocurran a los Oficiales Reales de Veracruz". AGNM, Industria y Comercio, vol. 10, exp. 11, f. 250v.

private agents, not only *asentistas*, provided that they met the requirements. In addition, Gálvez authorised the general attorney of Mexico to request a technician from Havana, whose salary was to be paid by the *caja real* of Veracruz, to supervise future felling operations. Finally, the viceroy asked for reports from the governor and the royal officials in the harbour of Veracruz, including price estimates and their view concerning the viability of the project.[25]

On 31 December 1783, a committee formed by the governor and the royal officials of Veracruz set out the price of good cedar to between 4 and a half and 5 pesos per cubic cubit delivered at the harbour.[26] In the first half of 1784, shipbuilding technician Don Luis del Toral was sent from Havana to Tlacotalpan to meet interested private contractors. Soon after, the first *asientos* were awarded to José Jiménez, Esteban Bejarano, Pedro Moscoso, and Ramón Carvallo,[27] who committed to deliver over 20,000 cubic cubits "of all kinds, from keels to stocks, according to the instructions and dimensions pointed out in the instructions set out by Don Luis del Toral".[28] An example of these guidelines is illustrated in Figure 17.

4   The *Asientos* Admitted in Veracruz (1784–1787)

The first four major *asientos*, granted to members of the Veracruz elite, confirmed the impressions of military engineer Miguel del Corral, who, during his survey of the region between the Alvarado and Coatzacoalcos rivers, found several areas with abundant wood that was suitable for naval construction. Apparently, del Corral's suggestions were taken into account for these contracts. The *asientos* were granted to José Jiménez, Militia Captain at Tuxpan, on the leeward coast, who committed to deliver 10,000 cubic cubits, including all sorts of parts in cedar and other hardwoods; Esteban Bejarano,[29] *vecino* of

---

25   Ibid., fs. 250–251.

26   Ibid., f. 253v.

27   Rafal Reichert "El transporte de maderas para los departamentos navales españoles en la segunda mitad del siglo XVIII", *Studia Historica. Historia Moderna* 43, no. 1 (2021), 59.

28   "todas clases de madera desde quillas hasta cepos de anclas con arreglo al plan formado por el delineador de construcción, don Luis del Toral, con arreglos a sus dimensiones señaladas y en los reglamentos que se les entregaron". AGNM, Industria y Comercio, vol. 31, exp. 5, f. 104v.

29   It is interesting to note that, according to the *Gaceta de México*, Esteban Bejarano was awarded a contract to build a brig in Tlacotalpan for 8,000 pesos de a ocho. The ship was launched on 21 February 1790 with the name *Nuevo Conde de Floridablanca*. Hemeroteca Nacional de México, *Gaceta de México*, 23 March 1790.

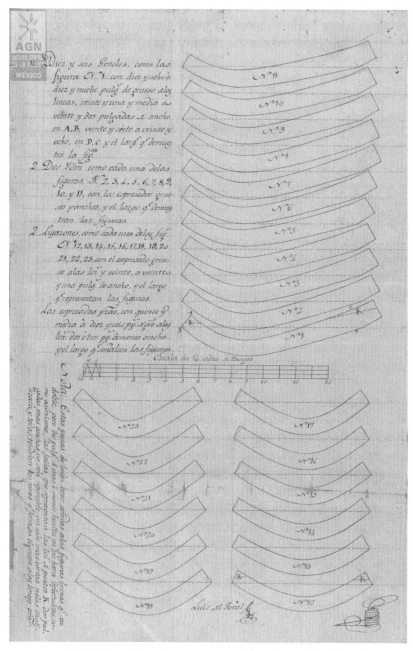

FIGURE 17   Wooden *genoles* for the construction of ships designed by del Toral with the
details of each carved part.

SOURCE: ARCHIVO GENERAL DE LA NACIÓN DE MÉXICO, MAPAS,
PLANOS E ILUSTRACIONES (280), MAPILU/210100/4485 GENOLES DE
MADERA CONSTRUCCIÓN DE BUQUE (4211)

Veracruz, whose order was for 2,000 cubic cubits of all sorts of parts and 50 knees of all varieties; Pedro Moscoso, resident of Acayucan, and Ramón Carvallo, *vecino* of Tlacotalpan, who committed to 4,000 cubic cubits of all sorts of parts in cedar.[30] When the *asientos* had run their course, Luis del Toral wrote that a large volume of wood had been delivered in Veracruz between 1784 and 1786, but also that not all targets had been met, as the contractors had not come forth with all the requested parts, which was causing delays in the Spanish shipyards.[31] For this reason, in 1786 Carvallo, Moscoso, and Bejarano received new contracts for 7,500 cubic cubits of wood, and other prepared pieces, and a new one was signed with Don Francisco Sánchez de Burgos, subdelegate at La Antigua, who committed to deliver 3,500 cubic cubits of all sorts of parts in cedar.[32]

The first *asiento* to be signed is of special interest, because the *asentista*, Don Ramón Carvallo, was already in possession of cedar trunks felled on the banks of the Papaloapan River and kept in storage in Tlacotalpan. The contract was signed by Don Joseph Serrano, Carvallo's representative in Veracruz, who committed to deliver 4,000 cubic cubits of naval parts. The contract does not mention felling, referring only to transport and delivery. The contract binds the contractor to deliver in Veracruz the cedar woods that Don Luis del Toral had seen, and approved of, in Tlacotalpan. Carvallo had to foot the transport bill, but the *Real Hacienda* covered taking the timber out of the water and putting it in storage, for which tasks the royal officials used forced labour. The contract indicates that this operation had to be undertaken on arrival and under the supervision of royal officials, including del Toral so that substandard pieces could be sent back. The *tozas* that passed the exam became the responsibility of the warehouse officer in Veracruz, and only then was payment to be made: 4 pesos and 5.5 reales. Carvallo received an advance payment of 8,000 pesos to organise the delivery and transport and pay the necessary wages. The contract imposed a 5% *alcabala* tax on the *asentista* at the end of the transaction. This *asiento* is unique, in that the contractor already had the wood at his disposal and was confident that he could sell it to the Crown, as was eventually the case.[33] Carvallo signed a second contract in August 1786, concerning dressed pieces, both boards and bends, of various sizes. The latter had to be between 6

---

30   Reichert, "¿Cómo España trató de recuperar su poderío naval?", 85.
31   AGNM, Industria y Comercio, vol. 10, exp. 11, f. 288.
32   Reichert, "¿Cómo España trató de recuperar su poderío naval?", 85.
33   AGNM, Industria y Comercio, vol. 10, exp. 11, fs. 254–259v.

and 7 *varas* long, so they could be used in the constructions of ships-of-the-line and frigates.[34]

In November 1787, the accountant of the *caja real* in Veracruz, Juan Matías de Lacunza, received a report from Luis del Toral concerning the wood that had been delivered at the harbour. The document notes that the pieces dressed by Carvallo had been dispatched to Cádiz (1,294 cubic cubits), Havana (190 cubic cubits), and El Ferrol (58 cubic cubits); the rest remained in Veracruz and some in Tlacotalpan. The cargoes setting sail to Cuba and Spain included various types of bends, clamps, *tozas* for planking, and boards.[35] According to the report, the contract expired in 1786, but after August no wood was sent to the shipyards owing to shortages in both navy and private shipping.[36]

The second *asiento* was awarded to Don José Jiménez, who, as noted, had been working with royal officials in Veracruz since the 1760s, first during the felling of wood for masting in Chimalapas and later when he signed a contract to supply wood for the renovation of the Castle of San Juan de Ulúa in 1775.[37] The contract signed in 1784 was more detailed than Carvallo's and is accompanied by a cover letter in which the contractor made a series of concessions. First, he offered a 40% discount on ordinary knees, although few such parts were included in the contract, which was for a total of 10,000 cubic cubits. He instructed that the trees were to be felled near the harbour of Tuxpan, where he lived, claiming that nobody else could cut down cedar or other woods suitable for naval construction in the area. The *asentista* suggested felling on both sides of the harbour and committed to dispatching the wood immediately after it was cut down. Interestingly, he committed to continue working even if war between European powers broke out. In addition, as a Militia Captain, he emphasised that his workmen and raftmen would take military training. On the other hand, he requested the support of the province's royal officials to recruit labour, which was, in his experience, not averse to running away after pocketing part of their wages. For this reason, he demanded that the governors of indigenous villages send him workers should he need them. The contractor mentions that efforts were to be made so that the parts complied with Don Luis del Toral's size requirements, but also requested permission to deliver them in other sizes, not to waste wood that was already available.[38] In this

---

34   AGNM, Edificios públicos, tomo 41, exp. 10, fs. 151–183v and AGS, SMA, Asientos, leg. 637.
      Copia de contrata aprobada por el virrey de Nueva España con don Ramón Carvallo.
35   AGNM, Industria y Comercio, vol. 31, exp. 8, f. 182.
36   AGNM, Edificios públicos, tomo 41, exp. 10, fs. 151–183v.
37   AGNM, Industria y Comercio, vol. 10, exp. 6, fs. 112–124v.
38   Ibid., exp. 11, fs. 261–264

document, the contractor demonstrates a good understanding of forestry, tree species, and woodworking. For instance:

> nobody doubts that, should everyone be allowed to choose the area freely [...] and should woodmen be allowed to cut Wood at their discretion, they will look for those Trees which are easiest to fell, because as [they] are for their own purposes, which do not need particularly hard woods, they will go for the tenderest or youngest ones, and it necessarily follows that, in a few years, no Wood suitable for Construction will be found in that area [of Tuxpan], forcing contractors to go farther into the hills, where the greatest difficulty exists in terms of transport, making it impossible for them to reap the profit that they expect.[39]

In another passage in his letter, Jiménez emphasised the need to protect woods suitable for naval construction, forbidding private persons and locals from felling them. This, he argues, was of the utmost importance for the Navy and the *Real Hacienda*, whose expenses inevitably increased with the additional cost of transport and shipping. He used Havana's example; so much indiscriminate tree cutting took place that to make the bottom of ships they had to begin bringing wood from elsewhere in Cuba and the Greater Caribbean, for instance Coatzacoalcos, Yucatan, New Orleans, and La Española, costing dear to the *Real Hacienda* in extraction, shipping, and transport costs.[40]

It is interesting to note that, on 20 November 1776, the king had to establish the *Junta de Maderas de Cuba*, whose creation was justified with the following argument:

> His Majesty, being aware of the deterioration of the cedar population in Cuba, caused by the excessive consumption of this material to make sugar crates; and given that no measure has been put in place to sort this problem, has decided to create in Havana a *junta*, with the participation

---

39   "ninguno duda que permitiéndose en los montes, que yo elija y prevea, que pueda hallar [...] madereros que se ocupen en cortar Maderas para particulares, estos buscarán aquellos Arboles, que les sean más fácil su derribo, porque como quieran que [son] para fines, que los solicitan y no necesitan de mayor robustez, se aplicarán a aquellos que sean más tiernos, o de poca edad, siguiéndose de esto necesariamente que dentro de poco años, ya no podrá hallarse con facilidad en aquel paraje [de Tuxpan] u otro que destine Maderas de Construcción siendo preciso quizás ocurrir a lo anterior de la Montaña a donde la mayor dificultad o mayor trabajo en el arrastre imposibilite al Contratista dar la Madera con aquella equidad que de presente tiene prometido". Ibid., f. 264v.

40   ibid., f. 265.

of the Navy commander, the governor, and the intendents, which, heeding if they will the advice of factory owners, will forbid the use of cedar for making crates, and order the use of an equivalent wood.[41]

The board was constituted by royal officials, including the governor, the navy commander, the navy commissar, and the Army and *Real Hacienda* intendents, as well as occasionally also by representatives of several political – economic sectors in the island. A conflict between the governor, Luis Unzaga, and the commander of the fleet and naval base, Juan Bautista Bonet, ensued because the former defended the interests of residents and major sugar producers, who needed much wood to pack sugar and burn in their boilers, while Bonet argued in favour of the interests of the *Marina Real*, hoping to restore Havana's prestige for shipbuilding while supplying Iberian departments with Cuban cedar, mahogany, and sabicú.[42]

Going back to Captain Jiménez's *asiento*, his concern for the forests of New Spain must be noted. He illustrated his point with the prohibition to cut oak, beech, holm oak, cork oak, and chestnut without authorisation, issued by Navy intendents in Spain in order to protect naval supplies. He claimed that his information about developments in Havana and Spain came from his collaboration with del Toral,[43] and he used it cunningly to forward his contract: he argued that should prohibitions not follow and tree felling to south of Veracruz come under control, the situation witnessed by the windward coast of the Gulf of Mexico would repeat itself: "as it was and is ransacked, there is no wood left to be found";[44] wood used for naval construction, particularly, was hard

---

41   "enterado S.M. del deterioro de cedros, que se experimentaba en la isla de Cuba por el excesivo consumo para cajas de azúcar; y no constando haberse verificado la providencia que para acabar con este inconveniente se resolvió que se formase en La Habana una junta integrada por el comandante de Marina, gobernador e intendentes, en que oyendo si les pareciese a algunos dueños de ingenios, acordasen la prohibición del uso del cedro proporcionando otra madera equivalente a la producción de cajas". Orden real de 23 de mayo de 1772 cited by Miguel Jordán Reyes, "La deforestación de la Isla de Cuba durante la dominación española (1492–1898)". PhD dissertation, Universidad Politécnica de Madrid, 2006, 54.

42   José Manuel Serrano Álvarez, "Élites y política en el astillero de La Habana durante el siglo XVIII", *Obradoiro de Historia Moderna* 28 (2019), 96–97.

43   As a Navy officer and responsible for measuring the wood, it is likely that del Toral knew about the ordinances published in January and December 1748, which were the basis of those published in 1762 and 1775. Pilar Pezzi Cristóbal, "Proteger para producir. La política forestal de los Borbones españoles", *Baética: Estudios de Historia Moderna y Contemporánea* 23 (2001), 583–596.

44   "por haber sido y ser transitada y saqueada ahora carece de madera".

to come by, owing to the reluctance of land owners, who tried to protect their forests.[45] Jiménez also made some observations about the cedar available in the south coast; in the former, cedars were of the male kind, and of the female in the latter. According to carpenters, the latter type was more resistant and durable. His proposal also mentions the purchase of timber by small ships, for their own repairs or for resale in Europe, which suggests that wood was being illegally traded on the coast of Mexico, as confirmed by some works on contraband. Finally, the *asentista* pointed out that virgin forests which could be used for shipbuilding could still be found in the jurisdictions of La Antigua, Nautla, Tampico, and other districts on the road to Nuevo Santander, in the windward coast.[46]

This contract and the attached proposals to limit tree felling on the south coast of the Gulf of Mexico were passed to the royal officials in Veracruz, Miguel del Corral, Francisco Agudo, Matías de Lacunza, and José María Lazos, who questioned Jiménez's monopoly over tree felling implied by his proposal. They, however, found his ideas concerning forestry management interesting. The committee insisted that it was important for wood only to be extracted at the time and place established by *asientos*, while forbidding private agents from cutting down young trees, especially concerning hardwoods such as cedar, mahogany, and sabicú.[47]

When José Jiménez's *asiento* was eventually approved in September 1784, del Toral gave him precise instructions, with drawings and samples (Figure 17). The document is divided into three sections: "hard wood", "straight wood" and "cedar wood". The first section described keels, stems, heads, fore sleepers, stemsons, stern post knees, cheeks, standards, keelsons, transoms, cat heads, large main posts, and timber of different lengths, widths, and thicknesses. For instance, concerning keels (see Table 46):

> these parts will have four flat faces; these dimensions, must not of necessity be applied to fresh edges throughout; even if the edges present some fail, as long as this can be fixed by shaving at least two inches off so that the new fresh edges meet the required dimensions, and not even in all faces, but in those in which rabbets will be open to fit the planking.[48]

---

45   AGNM, Industria y Comercio, vol. 10, exp. 11, f. 265.

46   Ibid., fs. 266v–267v.

47   Ibid., f. 269v.

48   "estas piezas tendrán sus cuatro caras rectas; las dimensiones que se señalan [Tabla 46], no deberán con precisión tenerlas a esquina viva en toda su extensión, pues con las expresadas dimensiones aunque por sus cantos tengan algunas fallas, no serán defectuosas, siempre que estas sean en tales términos que quitadas a la pieza dos pulgadas a lo menos

TABLE 46   Dimension of keels set out in Luis del Toral's specifications for José Jiménez's *asiento*

Keels	from 23 to 24 inches thick	from 25 to 26 inches wide	from 20 to 22 cubits long
Keels	from 16 to 17 inches thick	from 18 to 19 inches wide	from 17 to 18 cubits long
Aft keels	from 23 to 24 inches thick	from 25 to 26 inches wide on one head; 30 or more inches on the other if the tree allows	from 21 to 24 cubits long
Aft keels	from 16 to 17 inches thick	from 18 to 19 inches wide on one head and from 25 to 27 inches on the other	from 17 to 18 cubits long

SOURCE: AGNM, INDUSTRIA Y COMERCIO, VOL. 31, EXP. 3, FS. 65–65V

In this passage and others that deal with the dressing of the parts, del Corral left some room for imperfection, arguing that not all parts that were not pitch-perfect needed to be rejected. In this, he was particularly referring to knots in the wood, which could debilitate it. Similarly, although these specifications did not mention tree species, it must be assumed that most of the parts came in mahogany and sabicú, the American woods in which the pieces listed in the order were typically made.[49]

In the second section of del Toral's instructions, he referred to "straight woods" to make main posts, back of stern posts, aft sleepers, binnacles, rudders, crosspieces, capsterns, quarter decks and forecastles, *tozas*[50] for boards, sills, bowsprits, hold beams, sheet bitts, gunnel boards, quarter-deck boards, thick stuffs, ribbands, binding strakes, decks, head ledgers, tillers, and anchor stocks of different lengths, widths, and thickness. Like in the previous chapter, del Corral gave indications about how to dress the wood:

---

para dejarla en su perfección, pueda después quedar a esquina viva en toda su extensión, cuando no por todas sus caras precisamente por las en que deben abrirse los alefrices para recibir la tablazón". Ibid., vol. 31, exp. 3, fs. 65–68.

49   Gaspar de Aranda y Antón, "Las maderas de Indias", *Asclepio* 45, no. 1 (1993), 217–248; Reinaldo Funes Monzote, *From Rainforest to Cane Field in Cuba: An Environmental History since 1492* (Chapel Hill: University of North Carolina Press, 2008), 39–82; John T. Wing, *Roots of Empire: Forests and State Power in Early Modern Spain, c.1500–1750* (Leiden: Brill, 2015), 193–199.

50   Large block of mahogany or cedar, as big as each individual tree can provide.

– dressing gunnel boards, thick stuffs, and binding strakes, the thicker pieces will be as wide as possible, so that the thinner ones are as wide as practicable;
– with tillers, the longest will be also the thickest, so that thickness and length go in proportion;
– anchor stocks will be dressed on both sides.[51]

The final section, entitled "cedar wood", listed several parts, such as closed aft transoms, futtocks, *busardas*,[52] knees, curves for the base of heads, square knees, quarters, top timbers, *contra aletas*,[53] raised rungs, and horizontal rider rungs. Interestingly, del Toral recommended rungs to be made from trunks from which futtocks had been extracted first, in order to make the most of each trunk. Cedar was also used for *gambotas*[54] for the sides and the middle of aft, futtocks, first, second, and third ribs, kevel heads, buts, whole beams, stocks for counter or side timbers, head timbers, and planking.[55]

Del Toral also roughly evaluated the pieces as a price guide for the *asiento*. For instance, the felling and dressing of hardwood for keels was costed at 50 pesos; preparing a path and hauling trunks overland at 150 pesos,[56] and floating them downriver at 25 pesos; in terms of volume, keels fetched 8 pesos and 2 reales per cubic cubit. He also evaluated hardwood ribbands and thick stuffs at 52 pesos, or 5 pesos, and 1 real per cubic cubit. Concerning cedar, felling and dressing knees was priced at 3 pesos, overland hauling at 20 pesos, and floating downriver at 2 pesos; this amounted to 4 pesos and 5 reales per cubic cubit. Top timbers were costed at 29 pesos, 3 pesos, and 1 real per cubic cubit; *tozas* for planking were costed at 18 pesos, 2 pesos and 5 reales per cubic cubit. These price estimates were to be applied to all *asientos* signed after those granted to Carvallo, Moscoso, Bejarano, and Sánchez de Burgos.[57]

In a report dated 14 March 1785, after examining the wood delivered by José Jiménez to the royal warehouse in Veracruz, Luis del Toral acknowledged receipt of the following pieces:
– 1 aft closed stock;
– 3 futtocks;
– 5 first rungs;
– 11 second rungs;
– 6 third rungs;

51   AGNM, Industria y Comercio, vol. 31, exp. 3, fs. 68–69v.
52   Eng. Breast-hook.
53   Eng. Fashion-piece.
54   Eng. Gounter timber or head timber.
55   AGNM, Industria y Comercio, vol. 31, exp. 3, fs. 69v–77.
56   For a distance of 3 leagues.
57   AGNM, Industria y Comercio, vol. 10, exp. 11, fs. 282–283.

- 1 kevel rung;
- 4 butts;
- 6 buttocks;
- 5 knees for first;
- 1 curve for second;
- 3 knees for para quarter-deck;
- 2 pieces for head rails;
- 2 chains for counter or side timber stocks.

The size of these pieces was suitable for 60- and 70-gun ships-of-the-line. Del Toral also found the following frigate parts in the warehouse:

- 1 raised rung;
- 4 first rungs;
- 17 second rungs;
- 3 third rungs;
- 1 *aleta*;
- 8 kevel rungs;
- 2 butts;
- 6 pieces for buttocks;
- 3 beams for first;
- 4 second knees;
- 8 knees for quarter-deck;
- 3 for mess table knees.

Similarly, he counted 36 curved pieces for waterways, 216 straight pieces for sleepers, and 107 boards. In total, del Toral inventoried 470 prepared pieces, with a volume of 2,587 cubic cubits; however, he stressed how important it was to adhere to the prescribed sizes, because not all the parts delivered did so. For this cargo, the *asentista* was paid 10,000 pesos by the *Real Hacienda* in Veracruz.[58]

Finally, in his 1787 report, del Toral reports that of the wood delivered by José Jiménez, 1,263 cubic cubits of dressed hardwood and cedar were sent to Cádiz, 389 to El Ferrol, and barely 148 to Havana.[59] Like with Carvallo, the *asentista* met his targets in 1786, when the rest of the wood was stored in Tuxpan and Veracruz before being dispatched to Spain and Havana.[60]

In May 1785, the royal officials signed a contract with Esteban Bejarano for 4,000 cubic cubits of cedar from Acayucan, to be stored in Tlacotalpan. The contractor committed, first, to deposit 2,000 cubits of straight and curved *tozas* of different lengths and widths and 200 boards at a price, of 2 pesos and 2

---

58   Ibid., vol. 31, exp. 3, fs. 78–79.
59   Ibid., exp. 8, f. 182.
60   Ibid., exp. 3, f. 91.

reales, fixed by the *Real Hacienda*. Once the wood had been delivered, the contractor had to wait for Luis del Toral's inspection before sending it to Veracruz. In order to meet felling and transport costs, Bejarano requested a down payment of 3,000 pesos, with the guarantee of a merchant from Veracruz called Antonio Rodríguez Robledo. The *asiento* was approved by the royal officials in Veracruz Miguel del Corral, Francisco Agudo, Matías de Lacunza, and Vicente José de Mora.[61]

In the second contract, which was, in reality, an extension of the previous one, Bejarano committed to deliver 4,000 cubic cubits in two instalments of 2,000 cubits. The one-year contract was signed in late March 1786 and specified that the wood was to be sourced exclusively from the hills of Solcuautla, in the jurisdiction of Acayucan. This time, Bejarano gave assurances that the pieces would meet the detailed size specifications set out by del Toral: four keels for ships-of-the-line in *guapinol*, between 23 and 24 cubits long; one forefoot for a keel, between 18 and 20 cubits long; 100 standards for the first, second, and third-floor timbers (these made of cedar), of different lengths, between 4 and 8 cubits; *busardas* of different sizes, between 3½ and 8 cubits; futtocks between 10 and 12½ cubits long; ship-of-the-line rungs between 9½ and 11 cubits long, and for frigates, between 7½ and 8 cubits long,[62] and kevel rungs of different lengths (between 14 and 15 cubits for ships-of-the-line and between 7 and 7½ cubits for frigates).[63] In addition, del Toral ordered straight *tozas* over 12 cubits in length, and curved *tozas* at least 10 cubits long. Bejarano requested the price of waterways and sleepers to be adjusted to 32 reales per cubic cubit. The *asentista* also demanded a down payment of 8,000 pesos to begin work, with the same guarantor, Don Antonio Rodríguez Robledo, who was represented by Don Domingo Mirón. This second contract was approved by the junta of Veracruz on 28 March 1786.[64] It is interesting to note that the pieces delivered by Esteban Bejarano through these two contracts were sent only to the naval departments of Cádiz (1,437 cubic cubits) and El Ferrol (369 cubic cubits).[65]

Pedro Moscoso, Militia Captain in the province of Acayucan, signed his first contract in September 1785, aiming to deliver 4,000 cubic cubits of straight and curved *tozas* to the banks of the Michapa River, before they were sent to Veracruz, before the expiration of the agreement, ten months later. Moscoso

---

61   Ibid., vol. 10, exp. 14, fs. 416–426v.

62   In this case, the set price was 38 silver reales per cubic cubit. Ibid., vol. 31, exp. 6, f. 123v.

63   Del Toral also pointed out that, in order to reduce costs, double rungs could be made in the hills of Acayucan to be separated in the shipyard at a later date. Ibid., exp. 6, f. 144.

64   Ibid., fs. 122–155v.

65   Ibid., exp. 8, f. 182.

requested a down payment of 8,000 pesos with the guarantee of Don Antonio Fernández, *vecino* of Veracruz. It is worth noting that Moscoso, like the other *asentistas*, was a member of the Veracruz elite. Unfortunately, little is known about these characters, although Álvaro Alcántara López managed to locate some of them, including Pedro Moscoso, in the jurisdiction of Acayucan. Earlier in his life, this contractor had been hired by Don Joseph Quintero, the influential owner of the hacienda of Cuatotolapán and leading merchant in Coatzacoalcos. This alliance helped Moscoso's rapid promotion from treasurer to *alcalde mayor* in the province; later, he held the post of administrator of the salt and tobacco monopolies. In 1783, he helped his former patron in his dispute against the indigenous villages for the *realengo* lands of Acayucan, which he eventually added to his extensive estate.[66] This demonstrates the *asentista*'s economic and political leverage in the province and the business advantages that it brought with it. In fact, in his contracts with the *Real Hacienda* and the governor of Veracruz, Moscoso recruited indigenous workmen illegally.[67]

In this contract, Luis del Toral ordered 130 curved pieces for ships-of-the-line rungs, between 6 and 7 *varas* long, and 60 for frigates, between 4.5 and 6 *varas* long. The order also included waterways, buttocks, sleepers, and straight *tozería*, 8 *varas* long for 4 pesos and 2 reales per cubic cubit. This first *asiento* concluded with a partial delivery of the order to Veracruz while the rest were stored in the bar of Coatzacoalcos.[68] The second contract was signed in June 1786, and it concerned the felling of *guapinol* hardwood in a place called Guasuntan, near a stream that fed into the Coatzacoalcos. Moscoso was ordered keels, forefoots, heads, fore and aft sleepers, stemsons, stern posts, backs of stern posts, main stocks, knees, curves, keelsons, knight heads, cat heads, main pieces of rudder, and main capstern, boards for port cells or port stills, top timber lines, and quarter decks, at a price of 18 pesos per cubic cubit. In addition, the order included straight and curved *tozas* for beams and waterways for 4 pesos, curved pieces (futtocks, piques, crovy foots, *busardas*) and, unsurprisingly, planks for 6 pesos.[69]

In his report, del Toral wrote that he had exchanged information with Moscoso about a type of wood found near the Coatzacoalcos River, which he called "caobilla": "although this wood is not as solid as mahogany from Havana [...] it

---

66   Álvaro Alcántara López, "Redes sociales, prácticas de poder y recomposición familiar en la provincia de Acayucan, 1784–1802", in Antonio Ibarra and Guillermina del Valle Pavón (eds.), *Redes sociales e instituciones comerciales en el imperio español, siglos XVII a XIX* (Mexico: Instituto Mora, 2007), 236, 238–240.

67   AGNM, Justicia, tomo 183, exp. 28, fs. 134–136v.

68   AGNM, Edificios públicos, tomo 41, exp. 8, fs. 131–142.

69   AGNM, Industria y Comercio, vol. 31, exp. 5, fs. 103–109.

will be put to some of the uses to which this is put".[70] Del Toral did not authorise felling these trees under the contract, but requested several *tozas*, up to 2 *varas* long, which were sent to Havana to be inspected by the shipyard's carpenters. Moscoso's second *asiento* met greater difficulties in delivering the order, and according to del Corral's 1787 report, of the wood delivered in Veracruz, 1,952 cubic cubits were dispatched to Cádiz, 133 to El Ferrol, and 270 to Havana. Del Toral observed that 644 cubic cubits, part of the first contract, were still waiting in Coatzacoalcos; 109 cubits, delivered under the second contract, were sent to Cádiz, and 1,391 cubic cubits were still pending delivery.[71]

5   The Balance of the Wood *Asientos* Signed in New Spain (1784–1787)

In his November 1787 report, Luis del Toral estimated the volume of wood sent to the naval departments in 5,083 cubic cubits. He also reported that 1,037 pieces were stored in the royal warehouse in Veracruz (Table 47).

In the same document, del Toral specified how much wood was still pending delivery. Pedro Moscoso was 644 cubic cubits short from his first contract, and 1,261 cubits of hardwood and 6,000 cubits of cedar parts from the second; Esteban Bejarano still owed the *caja real* of Veracruz 2,427 cubits of cedar, from his second contract; Ramón Carvallo was 679 cubits of cedar short, also from his second contract; finally, Francisco Sánchez de Burgos had not yet delivered 3,500 cubits.[72] These figures show that the earliest *asientos* were a relative success because only in Moscoso's case they had fallen short of the target. The subsequent contracts, including the first one signed by Sánchez de Burgos,[73] all of which were signed in autumn 1786, were, however, more problematic, because the *tinglado* and the beach of Mocambo could not receive any more wood. In fact, on 31 January 1787, the royal officials of Veracruz were banned from signing any new contracts and asked to simply supervise existing ones. This was caused by the bottleneck created in the storage areas, but also by payment issues; the navy intendent of Havana's office, especially, was causing

---

70   [aunque esta madera no es de la solidez de la caoba de La Habana no obstante […] la
       aplicara no para todos pero para algunos fines que se emplea la recia]. Ibid., exp. 5, f. 109v.
71   Ibid., exp. 8, f. 182.
72   Ibid., f. 176.
73   This contractor made all preparations to begin felling, but at that point his contract was
       cancelled owing to the saturation of the temporary storages and the royal warehouse of
       Veracruz, which were full with the parts delivered by other *asentistas*.

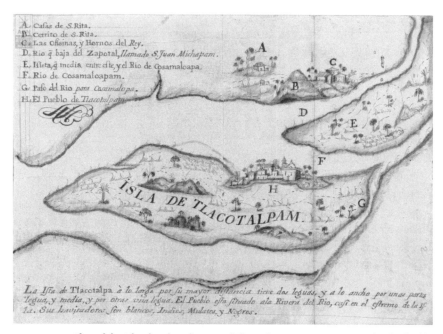

A. Casas de S. Rita.
B. Cerrito de S. Rita.
C. Las Ofisinas, y Hornos del Rey.
D. Rio q baja del Zapotal, llamado S. Juan Michapam.
E. Hleta, q media entre elle, y el Rio de Cosamaloapa.
F. Rio de Cosamaloapam.
G. Paso del Rio para Cosamaloapa.
H. El Pueblo de Tlacotalpam.

La Isla de Tlacotalpa à lo largo por su mayor distancia tiene dos leguas, y a lo ancho por unas partes legua, y media, y por otras una legua. El Pueblo esta situado ala Rivera del Rio, casi en el estremo de la Isla. Sus havitadores son blancos, Indios, Mulatos, y Negros.

FIGURE 18   Plan of the island and settlement of Tlacotalpan, an important site for shipbuild-
ing and logging in the province of Veracruz (1786)
SOURCE: ARCHIVO GENERAL DE LA NACIÓN DE MÉXICO, MAPAS, PLANOS E
ILUSTRACIONES (280), MAPILU/210100/2452 ISLA DE TLACOTALPAM, VER. (2353)

TABLE 47   Inventory of woods delivered by *asentistas* in Veracruz (1787)

Hardwood pieces	For ships-of-the-line	For frigates	Total
Forefoots	1	4	5
Keels	1	2	3
Stern posts		3	3
Main transoms		1	1
Heads		2	2
Keelsons	1	2	3
Pieces for timber lines	1	1	2
Straight ribbands	1		1
Bitts	1		1
Jear and top sail sheet bitts		2	2
Hatch beds		1	1
Anchor stocks	3	7	10
Tillers		1	1

TABLE 47   Inventory of woods delivered by *asentistas* in Veracruz (1787) (*cont.*)

Hardwood pieces	For ships-of-the-line	For frigates	Total
Total	9	26	35
Cedar pieces			
Double *busarda*		1	1
Keelson knees	1		1
Knees for the second deck	1	4	5
Knees for quarter-deck		4	4
Knees for mess tables		6	6
Rungs of all kinds, but not of different types	297	203	500
Curved sleepers			122
Butts	2		2
Buttocks	14	17	31
Straight *tozas* for planking			272
Boards			93
Total	315	235	1,037

SOURCE: AGNM, INDUSTRIA Y COMERCIO, VOL. 31, EXP. 8, F. 176

trouble because it was being charged for each cubic cubit of wood delivered to the naval department there.[74]

The wood business, which was controlled by the Veracruz elite, should have generated a profit in the region of 78,000 pesos de a ocho reales for a total of approximately 30,000 cubic cubits of prepared cedar and hardwoods. Storage issues in Coatzacoalcos, Tlacotalpan, and Veracruz, in addition to the difficulties posed by the scarcity of ships ready to bring the wood to Spain from Veracruz, forced the contracts to be suspended and, eventually, cancelled in November 1787. Standing obligations, in terms of both wood and money, were gradually settled, but some of the contractors, for instance, Bejarano and Sánchez de Burgos, were still requesting payment to offset their expenses in 1790, although the latter had, in the event, not delivered any wood at all.[75] Several parts for the 60- and 70-gun ships-of-the-line withered away in the sun and the rain, which are very detrimental for cut wood, on the beaches of Coatzacoalcos, Tlacotalpan, Tuxpan, and Veracruz. In February 1786, The *asentista*

---

74   AGNM, Industria y Comercio, vol. 10, exp. 11, fs. 320–323.
75   Reichert, "Recursos forestales, proyectos de extracción", 75.

Esteban Bejarano wrote that: "the damage that the wood suffers at the beaches, with the wind and the sun that they catch if they are not sheltered, is plain for all to see, [and] I commit to store it myself and at my own cost, until the king's ships arrive to the harbour [of Tlacotalpan] to take it to their final destination".[76] In view of this, the viceroy tried to make some use of these pieces and gave his authorisation to have them sold to private buyers, but this had little effect because the parts were conceived for ships-of-the-line and could not be fitted to build or repair smaller ships, like the ones used by merchants.[77] The contractors also offered their help, selling parts and planking to the private ships that arrived in Veracruz. In this way, José Jiménez sold baulks and beams to Don Antonio Sáenz de Santa María, who used it in construction, to be fitted as a rudder in the *San Pascual*, for 50 pesos, and a tiller to another ship for 23 pesos.[78]

The *asientos* signed in the province of Veracruz are a good illustration of the sort of initiative taken by navy officials in the viceroyalty after the American War of Independence (1775–1783). Following this victory over Great Britain, the Hispanic Empire pushed forth with its rearmament policies to improve its defences against the British threat.[79] In this context, the authorities in the colonies and the metropolis undertook a major project to supply cedar and

---

76   "es bien visible el daño y avería que reciben las maderas en la playa con los soles y vientos que cogen por falta de abrigo a efecto de evitarlo en esta parte, [y] me obligo a conservarlas en mi poder y de mi cuenta y riesgo hasta tanto que haya en el puerto [de Tlacotalpan] buques del rey que deban de conducirlas a su destino". AGNM, Industria y Comercio, vol. 31, exp. 6, f. 124v–125.

77   Andrade Muñoz, *Un mar de intereses*, 95.

78   AGNM, Industria y Comercio, vol. 31, exp. 3, fs. 86–87v.

79   John H. Parry, *The Spanish Seaborne Empire* (Berkeley: University of California Press, 1990); José Merino Navarro, *La armada española en el siglo XVIII* (Madrid: Fundación Universitaria Española, 1981); John Lynch, *Bourbon Spain, 1700–1808* (Oxford: Basil Blackwell, 1989); Grafenstein, *Nueva España en el Circuncaribe, 1779–1808*; Carlos Marichal, *La bancarrota del virreinato. Nueva España y las finanzas del Imperio español, 1780–1810* (Mexico: FCE, 1999); Stanley J. Stein and Barbara H. Stein, *Apogee of Empire Spain and New Spain in the Age of Charles III, 1759–1789* (Baltimore: Johns Hopkins University Press, 2003); Agustín Ramón Rodríguez González, *Trafalgar y el conflicto naval anglo-español del siglo XVIII* (San Sebastián: Actas, 2005); Juan José Sánchez Baena, Celia Chaín Navarro, and Lorena Martínez Solís (eds.), *Estudios de historia naval: actitudes y medios en la Real Armada del siglo XVIII* (Madrid: Ministerio de Defensa, 2011); Manuel-Reyes García Hurtado (ed.), *La Armada española en el siglo XVIII: ciencia, hombres y barcos* (Madrid: Silex Ediciones, 2012); Allan J. Kuethe and Kenneth J. Andrien, *The Spanish Atlantic World in the Eighteenth Century: War and the Bourbon Reforms, 1713–1796* (Cambridge: Cambridge University Press, 2014); Torres Sánchez, *Military Entrepreneurs and the Spanish Contractor State*; Juan Marchena Fernández and Justo Cuño (eds.), *Vientos de guerra apogeo y crisis de la Real Armada, 1750–1823*, vol. II (Madrid: Ediciones Doce Calles, 2018).

TABLE 48   Amounts of wood delivered in Veracruz and sent to the naval departments in 1784–1787,
according to del Toral's estimates

Asentistas	El Ferrol	Cádiz	La Habana	Veracruz	Total
		Cubic cubits and parts of 576			
José Jiménez	383 and 51	1,263 and 570	148 and 144	1,043	2,837
Esteban Bejarano,	369 and 474	1,437 and 138	0	941	2,747
second *asiento*	21 and 135	1,381 and 103	107 and 140	278	1,787[a]
Pedro Moscoso,	133 and 542	1,952 and 457	270 and 51	999	3,354
second *asiento*	0	109 and 150	0	81	190
Ramón Carvallo,	58 and 274	1,294 and 48	190 and 375	1,038	2,580
second *asiento*	61 and 374	2,391 and 251	164 and 554	702	3,318
Francisco Sánchez de Burgos	0	0	0	110	110
Total	1,034 and 122 cubic parts	9,829 and 568 cubic parts	881 and 113 cubic parts	5,193 and 198 cubic parts	16,938 and 427 cubic parts

a   Del Toral also listed 49 knees of all kinds.

SOURCE: AGNM, INDUSTRIA Y COMERCIO, VOL. 31, EXP. 8, F. 182

hardwoods for naval construction to the naval departments of Havana, Cádiz, and El Ferrol. In all probability, the initiative was inspired by the surveys carried out by Miguel del Corral and his assistants in several regions of Veracruz, Oaxaca, Acayucan, and Coatzacoalcos in the 1770s. However, as noted, the project fell short of its target: of the 30,000 cubic cubits expected from the seven *asientos*, only 11,745 cubits and 229 cubic parts arrived at their final destination, that is, the royal shipyards; 5,193 cubits and 198 parts were left in the warehouse of Veracruz and the beaches and riverbanks near the felling areas.[80]

Figure 19 shows the percentage of cedar and hardwoods dispatched to the naval departments of Cádiz, El Ferrol, and Havana during the operation of the *asientos* in New Spain. Cádiz received the lion's share of the wood provided by Veracruz from 1784 to 1787, with over 83%. This is unsurprising because Cádiz was the main port between Europe and America throughout the 18th century.[81]

---

80   AGNM, Industria y Comercio, vol. 31, exp. 8, f. 182.
81   Antonio García-Baquero González, *Cádiz y el Atlántico (1717–1778): el comercio colonial español bajo el monopolio gaditano* (Seville: Escuela de Estudios Hispano-Americanos,

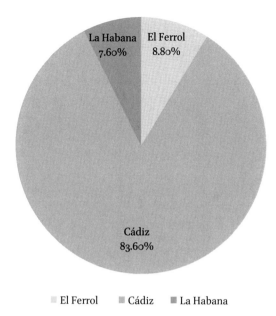

FIGURE 19   Arrival of New Spanish wood to naval departments (1784–1787)
SOURCE: AUTHOR'S OWN AFTER AGNM, INDUSTRIA Y
COMERCIO, VOL. 31, EXP. 8, F. 182

It may thus be assumed that these cargoes were later redistributed among
other naval departments, which is confirmed by the Navy intendent records in
El Ferrol and Cartagena, where mentions exist to the use of American wood.
For instance, on 18 March 1769, the department of Cádiz dispatched mahogany, anchors, and masting to Cartagena in a Dutch ketch. This responded to a
request posed by master shipbuilder Guillermo Turner in January. The mahogany parts included 12 pieces for pump brakes, 17 cubits long; 12 pieces for tillers, 15 cubits long; 24 pieces for mainmast and foremast beams, 9 cubits long;
12 pieces for mainmast and foremast crosstrees, 12 long; 12 pieces for mizzen

1976); Guadalupe Carrasco González, *Comerciantes y casas de negocios en Cádiz, 1650–1700*
(Cádiz: UCA, 1997); Ana Crespo Solana, "El patronato de la nación flamenca gaditana en
los siglos XVII y XVIII: trasfondo social y económico de una institución piadosa", *Studia
Histórica, Historia Moderna* 24 (2002), 297–329; Manuel Bustos Rodríguez, *Cádiz en el
sistema atlántico: la ciudad, sus comerciantes y la actividad mercantil (1650–1830)* (Madrid:
Sílex Ediciones, 2005); Francisco Javier Lomas Salmonte, Rafael Sánchez Saus, Manuel
Bustos, and José Luis Millán Chivite (eds.), *Historia de Cádiz: entre la leyenda y el olvido*
(Madrid: Editorial Sílex, 2005); Ana Crespo Solana, "Cádiz y el comercio de las Indias:
un paradigma del transnacionalismo económico y social (siglos XVI–XVIII)", *e-Spania* 25
(2016), 1–28.

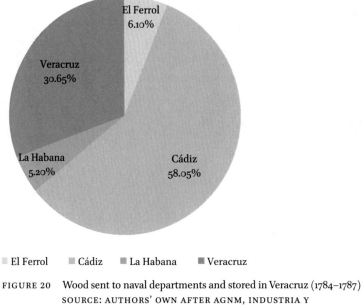

■ El Ferrol   ■ Cádiz   ■ La Habana   ■ Veracruz

FIGURE 20   Wood sent to naval departments and stored in Veracruz (1784–1787)
             SOURCE: AUTHORS' OWN AFTER AGNM, INDUSTRIA Y
             COMERCIO, VOL. 31, EXP. 8, F. 182

beams, 7 cubits long; and six crosstrees, 9 cubits long.[82] The another example, 29 February 1776, Cartagena-based officer Prisco Núñez wrote an inventory of the woods available, which included a column under the heading "Maderas de América", with 103 pieces and *tozas* of ácana, mahogany, and cedar, and even some of guayacán.[83]

The small presence of the Havana shipyard (7.6 %) is interesting, suggesting that up to 1787 Cuba's woodland reserves sufficed to meet the local naval construction demands; it must also be recalled that, on several occasions, the officials in Havana's naval department remarked that Cuban woods were better than the New Spanish ones.[84]

It has to be said that the attempt to use American timbers for naval construction, excluding Cuba, which was able to source much of its wood, did not reap as much reward as expected. This failure was mainly due to problems with transport, bringing the made pieces to the naval departments in Spain and Havana. The royal officials were aware of this problem and pointed out that,

---

82   AGS, SMA Marina, Asientos, leg. 343, Nota de los géneros que se facilitan de este arsenal [Cádiz] para el departamento de Cartagena.
83   Ibid., leg. 355, Relación de piezas de madera existentes en el arsenal de Cartagena.
84   AGNM, Correspondencia de Diversas Autoridades, vol. 41, exp. 67, fs. 316 y 316v.

owing to the lack of shipping, wood deliveries may have to wait until 1790.[85] This led to the loss of many cubits of wood, abandoned in the hills, temporary *tinglados* on the banks of the Papaloapan and Coatzacoalcos rivers, and the beaches of Veracruz. Following the losses for the *Real Hacienda* in New Spain, the Crown decided to cancel the project and seek its wood from elsewhere, such as New Orleans, which was used as a source for masting, and to increase the purchase of timber from northern Europe.

Finally, Map 10 illustrates the cabotage routes (in black) through which wood was transported from the mouths of the rivers to the main ports of the Gulf of Mexico, like Veracruz and Campeche. In these places, the timber was shipped in the king's ships, rarely in the private vessels that transported it to the naval departments in Spain and Havana (in yellow).

6   Wood Extraction Areas under the Jurisdiction of the Viceroyalty of New Spain in the Second Half of the 18th Century: Examples from Cuba and Louisiana

As noted, after the Spanish defeat in the Seven Years' War, the Crown launched several reformist projects to strengthen the army and the navy. In the previous pages, the projects and *asientos* unfolding in continental New Spain have been presented. Charles III, however, also promoted initiatives in other regions of the wide jurisdiction of New Spain. The most important programmes took place in the region of the Greater Caribbean, specifically in Cuba and northern Louisiana, where New Orleans became an important port for wood supplies.

7   Use of Cuban Wood by Havana's Shipbuilding Industry

As a result of its strategic location at the heart of the Spanish Empire in America, Cuba, especially Havana, turned into a key strategic position for the Spanish naval system of the *Carrera de Indias* in the 16th and 17th centuries,[86] in both commercial and military terms.[87] For this reason, an important

---

85   AGNM, Industria y Comercio, vol. 10, exp. 11, fs. 375–376.

86   From 1561 the fleets of New Spain and Tierra Firme converged in Havana for the return voyage to Spain. Clarence Henry Haring, *Trade and Navigation between Spain and the Indies in the Time of the Hapsburgs* (Cambridge, MA: Harvard University Press, 2014), 201–230.

87   Paul E. Hoffman, *The Spanish Crown and the Defense of the Caribbean, 1535–1585* (Baton Rouge: Louisiana State University Press, 1980); Allan J. Kuethe and José Manuel Serrano

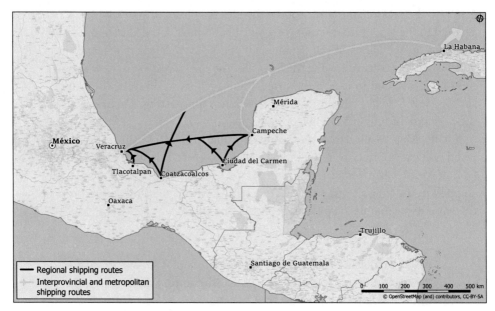

MAP 10   Maritime routes used to transport wood from logging sites to main ports and then to Havana
and Spain.
SOURCE: LOOK4GIS-LUKASZ BRYLAK BASED ON QGIS SOFTWARE

shipbuilding tradition developed in Havana beginning in the second half of
the 17th century, facilitated by the excellent local woods, at a time when 90
per cent of the island was covered in woodland.[88] Cuba was enormously rich
in wood, in a wide variety of species. The most highly valued for civil, military,
and naval construction were: cedar; sabicú; mahogany; poison ash; yaba; ocuje;
guayacán; quiebrahacha; guaraguao; jobo; mulberry tree; copperwood; maría;
jagua; majó; balsa; guaba; oak; and Caribbean pine.[89] In addition to being used
in ships, forts, houses, windows, doors, altars, and furniture, these woods were
also highly coveted by the sugar industry, which used them as fuel and to build
crates to pack the sugar.[90] It is important to recall that, from 1763 to 1792, Cuba

Alvarez, "El astillero de La Habana y Trafalgar", *Revista de Indias* 67, no. 241 (2007), 764–
765; José Manuel Serrano Álvarez, *El astillero de La Habana en el siglo XVIII: historia y
construcción naval (1700–1805)* (Madrid: Ministerio de Defensa, 2018), 20–39.

88   Funes Monzote, *Rainforest to Cane Field in Cuba*, 45.
89   Jordán Reyes, *La deforestación de la Isla de Cuba*, 23.
90   José Piqueras Arenas, "El azúcar en Cuba y las fuentes para su studio", *América Latina
en la Historia Económica* 6, no. 11 (1999), 35–47; José Jofré González, "The Sugar Industry, the Forests and the Cuban Energy Transition, from the Eighteenth Century to the

was the third largest cane sugar producer in the world, triggering the uncontrolled mass felling of virgin woods to expand sugar plantations. This drastically reduced the extension of the island's woodlands, essentially because of the abrasive cultivation methods in use and the systematic clearing of forested areas. To make matters worse, after two or three crops the newly cultivated land became sterilized and was abandoned, arresting the natural regeneration of wild plants.[91] This came on top of the ongoing struggle between sugar producers and the Navy for the use of wood. The problem became increasingly serious in the 1760s, prompting the organisation of the above-noted *Junta de Maderas de Cuba* in 1776. However, the conflict between the sugar industry and the *Marina Real* over the use of wood continued well into the 19th century.[92]

As noted, the root of the problem was the tug-of-war between sugar producers and the naval department in Havana, which in the early 18th century gained control of all woodland areas in the vicinity of its premises. This was confirmed by Act 13, Title 17, Book 4 of the *Recopilación de Indias* and by an order dated 8 April 1748, which established the source areas and types of wood to be used in the construction of the king's ships: cedar, yaba, ocuje, and guayacán, within 40 leagues to leeward and windward of Havana, 6 leagues inland from the northern coast, and 20 from the southern coast.[93]

Mass felling for naval construction began for the repair of the structure of the drydock in the shipyard of La Tenaza, destroyed during the British occupation of the harbour between August 1762 and April 1763.[94] This was the beginning of a new phase of ship construction in Havana. From 1766, when Gautier entered service in the Spanish Armada and Matthew Mullan relocated to Havana to take charge of the shipyard, the Navy introduced important changes in the construction of ships-of-the-line and frigates. The following year, Charles III ordered Lorenzo Montalvo Avellaneda, Count of Macuriges, intendent of the navy in Cuba, to build four new ships-of-the-line a year in La Tenaza. In

Mid-Twentieth Century", in César Yáñez and Albert Carreras (eds.), *The Economies of Latin America* (Cambridge: Cambridge University Press, 2014), 131–146.

91   Funes Monzote, *Rainforest to Cane Field in Cuba*, 82.

92   Reinaldo Funes Monzote, "Los conflictos por el acceso a la madera en La Habana: hacendados vs Marina" in José Piqueras (ed.) *Diez nuevas miradas de historia de Cuba* (Castellón: Publicacions de la Universitat Jaume l, 1998), 67–90; Serrano Álvarez, "Élites y política en el astillero de La Habana", 96–97.

93   Archivo Nacional de la República de Cuba (hereafter ANRC), Intendencia General de Hacienda, leg. 377, orden 2.

94   Elena A. Schneider, *The Occupation of Havana: War, Trade, and Slavery in the Atlantic World* (Chapel Hill: University of North Carolina Press, 2018), 221–222.

addition, the intendent, after consulting the fleet commander in Havana, Juan Antonio de la Colina, conceived the project of building a triple-decker ship.[95] The project crystallised in 1769 with the launch of the 112-gun *Santísima Trinidad*, which, after some later modifications, became the largest warship in the world.[96]

These political, economic, and military changes also affected wood sourcing by the Crown and the Navy. Wood *asientos* had been signed by members of the Havana commercial elite, which afterwards subcontracted the felling and transport operations, to the coast and, less often, to Havana itself, since the creation of the shipyard of La Tenaza in 1736. Between the 1760s and the 1780s, major contractors lost ground to smaller agents, such as Juan Blanco,[97] José Hilario Pérez, José de Aguiar, Miguel Oteiza, José de Miranda, Manuel Barranco, and Alejo Méndez, who began forming a new commercial elite in Havana through their participation in the wood business and their connections with the Count of Macuriges.[98] The *asiento* granted to Andrés Claro, who committed to supply guayacanes and mahogany pumps for the *Marina Real*, is a good illustration of the importance of these patronage networks. Initially, his proposal was rejected by the governor of Santiago de Cuba, Fernando Cagigal y García de Solís, II Marquis of Casa Cajigal, who was at loggerheads with Macuriges, but the latter's pressure, in both Cuba and Madrid (he argued that these woods were in short supply), finally overcame the governor's reluctance. He explained his support for Claro's proposal in the following terms: "the king's current need for these goods, and the orders that I have been given, force me to request, by all means necessary, this wood and all the rest that may be required for the construction of ships-of-the-line [...] the individual named above can deliver, and he commits to supply pumps and guayacanes to this shipyard for the established price".[99] The proposal was accepted in February 1767, and

---

95   Iván Valdez Bubnov, *Poder naval y modernización del Estado*, 372.

96   G.D. Inglis, "The Spanish Naval Shipyard at Havana in the Eighteenth Century", in *New Aspects of Naval History* (Baltimore: Department of History US Naval Academy, 1985), 47–58; José Manuel Serrano Álvarez, "La revitalización de La Habana en época de Lorenzo Montalvo, 1765–1772", *Revista de Historia Naval* 105 (2009), 71–101.

97   He was the main *asentista* for the arsenal of Havana from 1755 to 1762. AGS, SMA Marina, Asientos, leg. 624. El intendente de Marina de La Habana acompaña los autos originales [...] del asentista de maderas Juan Blanco.

98   Serrano Álvarez, "Élites y política en el astillero de La Habana", 88–90.

99   "la necesidad que el Rey tiene de estos géneros en la actualidad y las ordenes con que me hallo, me obligan a solicitar por todos medios esta madera, y otras de aquellas que convienen a la construcción de navíos; [...] el expresado individuo puede cumplir y se sujeta a proveer con abundancia este astillero de bombas y guayacanes, por los precios, y según las proporciones que se expresan".

afterwards Andrés Claro made several deliveries to Havana; each mahogany pump was between 18 and 21 cubits long and 18 inches thick, fetching a price of 14 pesos each; each *vara* of guayacán was between 10 and 20 inches in diameter, and fetched a price of 1.5 reales (the customary prices set out in wood *asientos*).[100] Claro's contract was still in force in April 1769, when a letter to Macuriges from Navy officer Antonio Esquerra reported the arrival in Havana of sloops *Señora de la Soledad*, captained by Antonio Roye, and *Nuestra Señora de Rosario*, by Miguel Gutiérrez, both of which were owned by Agustín del Risco, the *asentista's* partner. These ships were loaded with 200 *varas* of guayacán and 40 pump parts.[101]

It is worth pointing out that most wood resources for the Havana shipyard came from the hills of central Cuba, the provinces of Sancti Spíritus and Camagüey, the province of Holguín, the region of Manzanillo, and the bay of Jagua, in the south and south-east of the island. These resources allowed the shipyard to build ships cheaper and faster because the raw materials were closer to the workshops.[102] As noted, mass felling for the Navy began in 1766. For instance, the royal officials organised a large-scale operation in the area of Jagua, which, according to the Armada officers, had enough cedar, mahogany, and sabicú to build 60 ships-of-the-line in Havana and the Iberian shipyards. The region was thinly populated, so Macuriges and Colina decided to send a military officer to lead the expedition, along with a foreman, carpenters, and 50 pairs of oxen with their respective drivers. The operation was costed at 40,000 pesos. The project was endorsed by the Secretary of the Navy, Julián de Arriaga, on 18 August 1767, and the felling and first deliveries of wood in Navy ships began a few months later.[103]

Two years later, the Navy intendent ordered the survey and felling of 24,000 pieces of all kinds for the construction of 60- to 120-gun ships-of-the-line. Commander General Juan Antonio de la Colina, in his letters to Viceroy Croix in late 1769, mentioned that everything was ready to begin work, except for money, which was lacking.[104] The problem was, therefore, a shortage of funds and not of shipping, in contrast with New Spain. The department of Havana, despite having its own ships, developed extensive connections with local private skippers, commanding sloops, schooners, and even barges and bongos, which were actively involved in hauling wood; for instance, between 1766 and

---

100 ANRC, Intendencia General de Hacienda, leg. 20, orden 60.
101 Ibid., leg. 17, orden 96.
102 Reichert, "El transporte de maderas para los departamentos navales", 56.
103 AGNM, Marina, tomo 26, fs. 169–170 and 228–229.
104 AGNM, Correspondencia de Diversas Autoridades, vol. 13, exp. 44, fs. 202–203.

1770, captains Mario José Germosilla, Julián Rodríguez,[105] Joseph Castaña, Amaro López, Simón García, Esteban Rodríguez, Gregorio Melo,[106] and Juan Sierra signed contracts to transport wood for the Navy.[107]

8   Garibaldo's Project as the Model for State-Sponsored Exploitation of Wood in Cuba

Concerning the direct exploitation of Cuban wood by initiative of royal officials deployed in Havana, the project *Instrucción hecha por el contador de navío de la Real Armada, don Manuel Carlos Garibaldo para plantificar y establecer dos cortes de madera dura y de cedro con sus respectivos requisitos para construcción de dos navíos de 60 cañones*[108] was handed to Viceroy Don Antonio María de Bucareli y Ursúa in 1777. In this project, the Navy accountant presented a programme for the supply of different types of wood for the construction of ships-of-the-line in the Havana shipyard. It is important to note that Don Manuel Carlos Garibaldo was a Cuban Creole who, between 1767 and 1770, made several visits to royal felling areas where the Navy sourced ocuje, yaba, mahogany, sabicú, cedar, and other species, and supervised the shipment of the cut wood to Yaguajay (now the National Park of Caguanes), in the province of Sancti Spíritus, and thence to the shipyards in Havana. The felling areas were located in the jurisdictions of Cayayues, Charco Hondo, Guanabanabo, and Yaguajay.[109]

Garibaldo's experience in the field and later as accountant for the Navy intendent allowed him to prepare a detailed *Instrucción*, divided into 42 points in three sections, which dealt with:

– supplies such as nails, tools, animals, and food, as well as a chaplain and a surgeon; he also argued that a chapel and two barracks had to be built;
– personnel and wages, mentioning the use of inmates in Havana prison as a potential source of labour;
– exact measurements and amounts of hardwood (mahogany, sabicú, and yaba) and cedar required for two 60-gun ships-of-the-line.[110]

---

105   ANRC, Correspondencia de los Capitanes Generales, leg. 26, ordenes 36 and 149.
106   Ibid., leg. 27, ordenes 57, 83, 130, and 142
107   Ibid, leg. 30, orden 25.
108   AGNM, Indiferente virreinal, Marina, caja 4737, exp. 46, fs. 1–34v.
109   José Martínez-Fortún y Foyo and Humberto Arnáez y Rodríguez, *Diccionario biográfico remediano* (Habana: Ayuntamiento de San Juan de los Remedios, 1960), 56–57
110   AGNM, Indiferente virreinal, Marina, caja 4737, exp. 46, fs. 4–34.

Some points are worth examining in detail. They illustrate the organisation of felling operations and can be used as a model for the way wood for naval construction was sourced in America. First, Garibaldo noted that it was important to set out a timeframe for the operation, which he estimated in 12–13 months in total (felling, dressing on-site, and hauling to the rivers). He also recommended felling to begin between November and February, "to avoid the juices of said trees (cedar) making them rot later".[111]

Garibaldo made precise calculations of the wood requirements for the construction of two 60-gun ships-of-the-line: 3,877 parts for construction and 2,080 for planking, lining, and freeboard of cedar. He also suggested cutting one-third above requirements, for spare parts and to support construction works. He estimated that two ships-of-the-line would take 7,939 cedars and that in order for the works to be finished in 12–13 months 462 cedars should be felled every month.[112] In addition, these ships required 1,048 pieces of hardwood (sabicú, yaba, and mahogany); to provide for possible damage during transport, he increased the number to 1,223 trees, i.e. 88 pieces per month.[113] Once the trees had been felled, they would be dressed to make specific parts (hardwoods: keels, stern posts, back of the stern post, posts, feet, and other parts; cedar: planes, piques, futtocks, rungs, and knees). All tasks were to be supervised by the construction assistant and the foremen responsible for keeping all parts within the specified dimensions (length, thickness, and width).[114]

The accountant also spelled out the duties of each worker; the principal commissar was responsible for:

> cedars not to be felled outside the months of November, December, January, and February, no thinner nor narrower than half a *vara*. Each foreman will receive a list of the number of trees of each kind to be felled, and the parts and types that are required for the construction of two 60-gun ships-of-the-line so that they can keep on top of things punctually and no part is missing. Each month he must carry out an inventory of parts, oxen drivers, and the rest of the staff [...]. Accounts must be clearly and punctually kept [...] for the distribution of money to be kept abreast of production so that every four months they can report to their superiors

---

111   "no se experimente jugosidad competente de dichos arboles (cedro) para el logro de su sanidad y precaución de toda posterior pudrición". Ibid., f. 4.
112   Ibid., fs. 4v and 6.
113   Ibid., fs. 5 and 6v.
114   Ibid., f. 5v.

and so that the wood can be priced accordingly to the contract signed in this harbour by Francisco Franquis de Alfaro.[115]

In addition, the principal commissar was in charge of public order and preventing drunkenness, having the power to punish those who did not follow the rules.

Point 30 deals with assistants and foremen. They were responsible for the felling of trees in their assigned sectors and for informing their superiors of the number of trees felled. In addition, they were to supply the technical foremen with plans, figures, dimensions, and descriptions of each piece, following specifications. They were also to survey their surroundings regularly, to ensure that the trees were sufficiently thick, wide, and long, taking all necessary measures and supervising the work of foremen and carpenters to assess whether they were really earning their wages or not. They were to oversee the monthly delivery of dressed parts to the technical foremen, branding the parts as the king's with a stamp marked "VR". They also used a chisel to mark what part of the ship each piece was intended for and to confirm delivery to the oxen drivers, who at that point became responsible for the safety of the wood.[116]

Garibaldo also described the duties of the technical foremen and their assistants, who were to support the hands felling the trees and give priority to the pieces for which greater need existed according to the orders and plans

---

115   no se tumben árboles de madera de cedro en otras menguantes que las de noviembre y
        diciembre, enero y febrero de cada año y que su grueso y ancho no baje de media vara.
        Que a cada capataz delineador se le entregara una relación que exprese el total número
        de árboles de una y otra madera que deban derribar y el de las piezas por clases que son
        precisas para la construcción de dos navíos de 60 cañones a fin que puedan arreglar la
        labor de ellas en términos que al año o antes la verifiquen sin que falte una. Ha de pasar
        revista general cada un mes a la maestranza, los boyeros y demás operarios que estén
        empleados en dichos cortes previstos [...] Debe llevar con cuenta claridad y prolijidad
        que da [...] la respectiva distribución de caudales y de lo que vaya produciendo la labor y
        el tiro de las maderas a efecto de formar un tanteo cada cuatro meses y pasarlo a la supe-
        rioridad para la mejor inteligencia y que se comprueben los buenos o malos efectos que
        resulten reglados a los precios de la madera por los que expresa la contrata que celebró en
        este puerto don Francisco Franquis de Alfaro. (Ibid., fs. 16–17v)
            Francisco Franquis de Alfaro was regidor in Havana and one of the Army and the Na-
        vy's leading creditors. It is estimated that, between 1765 and 1800, he loaned 4 million
        pesos to these institutions. In exchange, he was granted *asientos* for the supply of wood,
        iron, copper, and other materials. José Manuel Serrano Álvarez, "Contratas militares en La
        Habana durante el siglo XVIII: riqueza local y visión imperial", in *Memoria de la Terceras
        Jornadas de Historia Económica* (Mexico: Asociación Mexicana de Historia Económi-
        ca-amhe, 2015), 379.
116   AGNM, Indiferente virreinal, Marina, caja 4737, exp. 46, fs. 17v–19.

supplied by the building assistant. Technical foremen were also to oversee the work of carpenters and made sure that no frivolous excuses were given to delay work. Every night, after dinner, they were to examine the pieces prepared in the course of the day and mark each part. At the end of the month, they presented a report to the principal commissar and the building assistant.[117]

Garibaldo also set out the wages to be paid to each member of staff. Technical foremen were to be paid 16 reales per day; their assistants, 12 reales; cedar carpenters, 7 reales (and had to buy their own axes and machetes); hardwood carpenters, 9 reales (the *Real Hacienda* customarily provided their tools: axes and machetes); road foremen, 13 pesos per month, and *caminantes*, 12 pesos; callipers, 8 pesos; quartermasters (responsible for food supplies), 14 pesos; cooks, 12 pesos; oxen foremen, 35 pesos; their assistants, 25 pesos; smart hands, 14 pesos, like *ramajeros* and *recogedores*—if they were forced workers, only 1 peso per month. Workers were given shoes, 'rain boots', and cloth tunics twice a year. Two barracks for the workers were to be built, including a kitchen and pantry; another one outside the complex, for carpenters and *caminantes*, and another one for ox drivers.[118]

Finally, in his last point, Garibaldo presented a series of tables with the pieces of cedar and hardwood required for the two 60-gun ships-of-the-line. In addition to the description of each part, these tables included the number needed for each, and the breakdown of prices for on-site work and hauling, as well as the total price. The prices suggested by Garibaldo follow Don Francisco Franquis de Alfaro's above-noted contract (Table 49).

TABLE 49   Wood parts required for two 60-gun ships-of-the-line in the Havana shipyard, according to Manuel Carlos Garibaldo's 1777 proposal

		Hardwoods			
Carved piece	Number	Price for on-site work [reales]	Price for hauling [reales]	Price per piece [reales]	Total [reales]
Aft keels	2	240	400	640	1,280.00
Fore keels	2	240	400	640	1,280.00
Straight keels	6	240	320	560	3,360.00
Stern posts	2	160	240	400	800.00

117   Ibid., fs. 23v–25v.
118   Ibid., fs. 9v–10.

TABLE 49 Wood parts required for two 60-gun ships-of-the-line in the Havana shipyard (*cont.*)

		Hardwoods			
Carved piece	Number	Price for on-site work [reales]	Price for hauling [reales]	Price per piece [reales]	Total [reales]
Back of the stern post	2	80	160	240	480.00
Feet	2	240	320	560	1,120.00
Heads or second feet	2	160	240	400	800.00
First inner posts	2	160	240	400	800.00
Second inner posts	2	160	240	400	800.00
Transoms	2	160	240	400	800.00
Standards or knees by that name	2	80	120	200	400.00
Aft sleepers	4	160	176	336	1,344.00
Sobre durmientes	4	160	176	336	1,344.00
Keelsons	8	160	320	480	3,840.00
Bitts	8	48	96	144	1,152.00
Cat heads	4	40	64	104	416.00
Apóstoles	8	60	176	336	2,668.00
Bolsters	8	32	80	112	896.00
Band knees	12	48	80	128	1,536.00
Head knees	2	32	64	96	192.00
Curved crossed ribbands	50	32	80	112	5,600.00
Straight crossed ribbands	100	32	64	96	9,600.00
Cintas de choza de vuelta[a]	16	32	48	80	1,280.00
Cintas de choza derechas	100	24	40	64	6,400.00
Carlings	80	24	40	64	5,120.00
Thick stuff	250	20	30	50	12,500.00
Oblique hold props	40	16	24	40	1,600.00
Straight hold props	40	16	24	40	1,600.00
Main piece of the rudder	2	80	176	256	512.00
Mahogany pumps, 17 cubits	8	64	120	184	1,472.00
Barrel of main capstern	2	32	64	96	192.00
Base of deck and quarter-deck	4	20	36	56	224.00
Beds, hatches and y ledges	160	12	16	28	4,480.00
Main and fore bitts	8	32	64	96	768.00

a   Eng. Wale.

TABLE 49   Wood parts required for two 60-gun ships-of-the-line in the Havana shipyard (*cont.*)

		Hardwoods			
Carved piece	Number	Price for on-site work [reales]	Price for hauling [reales]	Price per piece [reales]	Total [reales]
Legs of main beams	20	16	24	40	800.00
Mizzen legs	8	16	16	32	256.00
Stocks of first anchors	6	32	52	84	504.00
Stocks of second anchors	4	32	52	84	336.00
Stocks of third anchors	8	32	52	84	672.00
Sabicú pieces for port cells	32	64	160	224	7,168.00
Transom knees	8	80	120	200	1,600.00
Standards knees	2	80	120	200	400.00
Puercas	16	48	80	128	2,048.00
Total prepared pieces	1,048.00				

		Cedar wood			
Lying planes	90	24	32	56	5,040.00
Raised planes	30	24	32	56	1,680.00
Futtocks	26	24	40	64	1,664.00
Piques	12	24	40	64	768.00
Futtocks	400	16	24	40	16,000.00
Second rungs	400	16	24	40	16,000.00
Third rungs	400	16	24	40	16,000.00
Fourth rungs	400	16	24	40	16,000.00
Kevel rungs	400	16	24	40	16,000.00
Busardas	20	40	72	112	2,240.00
Horcazes para popa y proa	7	40	72	112	784.00
Contra transoms	6	32	48	80	480.00
Chains	6	32	64	96	576.00
Bolsters	24	16	40	56	1,344.00
Curved waterways	16	20	32	52	832.00
Straight waterways	32	32	64	96	3,072.00
Quarter-deck waterways	16	36	56	92	1,472.00
Curved sleepers	24	24	56	80	1,920.00
Straight sleepers	24	32	80	112	2,688.00
Piezas para zafrán	2	32	48	80	160.00

TABLE 49   Wood parts required for two 60-gun ships-of-the-line in the Havana shipyard (*cont.*)

		Cedar wood			
Carved piece	**Number**	**Price for on-site work [reales]**	**Price for hauling [reales]**	**Price per piece [reales]**	**Total [reales]**
Anguilas	12	32	48	80	960.00
Main deck beams	72	48	80	128	9,216.00
Middle deck beams	78	44	76	120	9,360.00
Baos de aire	46	44	76	120	5,520.00
Quarter-deck beams	48	36	60	96	4,608.00
Latas para toldilla	36	24	32	56	2,016.00
Main deck knees	348	24	27	51	17,748.00
Lodging knees	144	16	32	48	6,912.00
Open knees for middle deck beams	120	16	24	40	4,800.00
Knees for *llaves*	120	16	32	48	5,760.00
Knees for *baos de aire*	120	16	32	48	5,760.00
Quarter-deck beam knees	120	16	32	48	5,760.00
Transom knees	4	16	32	48	192.00
Chain knees	24	16	32	48	1,152.00
Counter timbers	20	16	40	56	1,120.00
Jibbooms	8	32	64	96	768.00
Contra aletas	8	24	32	56	448.00
Beak of cut waters	2	120	184	304	608.00
Piezas para leones	4	36	96	132	528.00
Pieces for buttocks	200	16	40	56	11,200.00
Pieces for finishings and quarter pieces	8	24	64	88	704.00
Total prepared pieces	3,877				

		Tozas			
Cedar, 9 *varas* long, ¾ wide, and ⅔ thick	400	32	84	116	46,400.00
Cedar 9 *varas* long, ⅔ wide and ½ *vara* thick	400	24	64	88	35,200.00
Cedar, ⅔ in square	400	24	64	88	35,200.00

TABLE 49   Wood parts required for two 60-gun ships-of-the-line in the Havana shipyard (*cont.*)

| | | *Tozas* | | | |
Carved piece	Number	Price for on-site work [reales]	Price for hauling [reales]	Price per piece [reales]	Total [reales]
Cedar, ½ *vara* thick in square	500	16	48	64	32,000.00
Cedar, 1 *vara* wide and ¾ thick	100	48	94	142	14,200.00
Total *tozas*	1,800				163,000.00
		Curved *tozas*			
Cedar, ¾ wide and ⅔ thick	80	32	84	116	9,280.00
Cedar, ⅔ wide and ½ thick	100	24	64	88	8,800.00
Cedar, ½ *vara* in square	100	16	48	64	6,400.00
Total curved *tozas*	280				24,480.00
		Financial summary			
Hardwood [reales]					90,440.00
Cedar wood [reales]					387,340.00
TOTAL [reales]					477,780.00
The Final Price 59,722.50 pesos. A 60-gun ship-of-the-line costs 29,861.25 pesos					

ᵃ Eng. Wale.

SOURCE: AGNM, INDIFERENTE VIRREINAL, MARINA, CAJA 4737, EXP. 46, FS. 33–34

The proposal was analysed by Antonio María de Bucareli in early 1778 but was rejected because he did not want to allocate extraordinary funds to undertake public-funded felling operations in Cuba, as a number of private *asientos* were already underway.[119] Although the project did not go further, Garibaldo's proposal can be used as a guide for the organisation of felling operations, especially those carried out by the Navy. The *Instrucción* gives a clear idea of the logistical, financial, and labour challenges posed by these enterprises in Cuba and perhaps also in other American regions and Spain.

---

119   AGNM, Indiferente Virreinal, Marina, caja 4737, exp. 38, f. 2.

Despite the fact that no new ships were built during the American War of Independence (1779–1783), the period 1766–1789 was the most active for the Havana shipyards in the 18th century. The most significant among the many war and transport ships built during this period were several triple-deck ships-of-the-line, including the above-noted *Santísima Trinidad* (1769) and, in the 1780s, all of this ships had 112-guns.[120] This activity was framed by constant conflict with sugar landowners and internecine personal rivalries within the Navy. For instance, the replacement of Count Macuriges by Commander Bonet in 1772 triggered mutual accusations of misgovernment and complicity with the sugar lobby. Lorenzo Montalvo Avellaneda, for his part, wrote a document to defend his performance as intendent of Havana, claiming that 15 ships of different kinds and eight brigantines had been built under his supervision: he also boasted of repairing 74 ships of different sorts, and of sending 7,000 *tozas* of cedar and mahogany and 27,000 of guayacán to Spain.[121]

Despite these results, naval construction began to decline, both in Spain and in Havana, in the 1790s, in all probability as a result of financial difficulties derived from the enormous cost of keeping the Crown's war machine in good order. In fact, several initiatives were put forward in the second half of the 18th century to find donations or credit to help cover these expenses. Between 1765 and 1799, the Cuban elite loaned 3,969,275 pesos to the *Real Hacienda* to fund naval construction in Havana.[122] Similarly, in 1776, José de Gálvez asked the Viceroy Antonio María de Bucareli to organise a round of donations to increase the "number of warships in Spain and the West Indies".[123] By this initiative, the king's subjects in New Spain collected 1,252,209 pesos in 1776 and 1777. Other significant contributions came from the *Tribunal de Minería* (300,000 pesos), the *Consulado de México* (300,000), and the Count of Regla (200,000).[124]

The 1790s witnessed the end of Havana and Cuba's central contribution to the resurgence of the Spanish Navy. The last of the major ships-of-the-line, the 112-guns *Príncipe de Asturias*, was launched in 1793. Afterwards, geopolitical conditions in Europe and America, naval war with Great Britain, and the crisis in the monarchy caused by Charles III's death in 1788, as well as lack of funds,

---

120   Serrano Álvarez, *El astillero de La Habana en el siglo XVIII*, 530.
121   Valdez Bubnov, *Poder naval y modernización del Estado*, 383–385.
122   Serrano Álvarez, "Contratas militares en La Habana durante el siglo XVIII", 379.
123   "número de embarcaciones de guerra en la península y en Indias".
124   Gillermina del Valle Pavón, *Donativos, préstamos y privilegios* (Mexico: Instituto Mora, 2016), 46–47.

spelled the end of the erstwhile protagonist role played by the Havana ship-yard for the Hispanic Empire.[125]

9   Illegal British Timber-Felling Operations in Cuba

Illegal felling was a problem shared by all woodland areas in the Greater Caribbean, including such an important military enclave for Spain as Cuba. Most of the "woodland pirates" active in Cuba were British groups operating in the southern coast—regions of Manzanillo, Portillo, Macaca, and Ocujal—from their bases in Jamaica. After carrying out an inspection in these regions in October 1766, Lieutenant Bernabé Zubieta reported to the Marquis of Casa Cagigal, that he had seen "several English [sic] armed ships, [smuggling wood] in the coasts of Manzanillo and other regions",[126] as well as "felling trees with armed people on land".[127] Zubieta claimed that this was a common occurrence, especially in the hills of Manzanillo, and that this damaged the forests, making them unusable for the *Marina Real*.[128] The British had chosen this area because 3 leagues of thick woodland—rich in mahogany, cedar, and brasilete—separated the coastal sand bar and the interior plain of Jarra; the forests, which in some areas began barely 70 or 80 *varas* from the coastline, had often been cleared with fire. From land, the illegal felling areas were visible from Cape Escalera, to leeward, to the mouth of the Gua River: between these areas and the Cove of Manzanillo, the smugglers' sloops could moor without difficulty. The governor of Santiago de Cuba, 2nd Marquis of Casa Cajigal, suggested the capital of the jurisdiction—Bayamo—to arrange a watch, and also for warships to patrol the sea. The governor pointed out that Lieutenant Bernabé Zubieta had tried to ambush the smugglers, but these attempts failed because the British had informants in the region, Spanish subjects who kept the British abreast of the presence of Spanish troops or militia. Zubieta reported that, when he arrived with his men "[the British] mocked us, setting sail to fell and smuggle elsewhere to either leeward or windward".[129] In January 1767, the 2nd Marquis of Casa Cagigal shared these reports with

---

125   Serrano Álvarez, *El astillero de La Habana en el siglo XVIII*, 500–526.
126   "diferentes embarcaciones inglesas armadas en las costas de los surgideros de Manzanillo y otras".
127   "hacían cortes de maderas, poniendo gente en tierra armadas".
128   ANRC, Correspondencia de los Capitanes Generales, leg. 15, orden 55.
129   "[los ingleses] se hallaron burlando, porque la embarcación se puso a la vela para hacer su corte y contrabando un poco más a Sotavento o a Barlovento". ANRC, Correspondencia de los Capitanes Generales, leg. 22, orden 27.

the Marquis of Croix, asking for financial assistance to pursue the smugglers in Manzanillo. The viceroy, however, rejected the petition, arguing that Cuba already received large funds for the Army and Navy. In consequence, the only reaction to the illegal felling operations was the permanent deployment of four militiamen and one captain in the bay.[130] The situation remained the same until 1779, when Spain entered the American War of Independence and the widespread hostility towards the British brought these operations to an end, at least on a large scale. In the 1780s, British wood smugglers found new areas, in Walix (modern Belize), to illegally source hardwoods for naval construction and, especially, furniture.[131]

<h2>10   The Wood Business in Louisiana and New Orleans in the Late 18th Century</h2>

Since they began building ships for the king, the shipyards in Havana faced wood supply issues, notably for ships-of-the-line. Cuban tropical forests yielded guayacanes and marías, but these were not the most suitable tree species, especially for mainmasts, foremasts, mizzens, and bowsprits. The right species could be found in Sierra Madre Oriental, in the provinces of Veracruz and Oaxaca, and, as noted, large felling operations were carried out in the 1760s and 1770s in the hills of Chimalapas; most of the pines felled in this region were dispatched to the shipyard of La Tenaza. However, as a result of the high felling and storage costs, as well as shipping shortages, these operations were cancelled in 1773,[132] when the Navy intendents in Havana took an interest in Louisiana and the harbour of New Orleans as an alternative source of wood. The earliest reference to the use of masting from this region is dated to July

130   AGNM, Correspondencia de Diversas Autoridades, vol. 11, exp. 32, fs. 156–158.
131   Samuel J. Record and Robert W. Hess, *Timbers of the New World* (New York: Arno Pr,1972), 236–245; Adam Bowett, "The Jamaica Trade: Gillow and the Use of Mahogany in the Eighteenth Century," *Regional Furniture* 12 (1988), 37–39; Adam Bowett, "The English Mahogany Trade 1700–1793" (London: Brunel University, 1996, [PhD dissertation]), 90–94, 132–137, 152–192; Michael Camille, "The Effects of Timber Haulage Improvements on Mahogany Extraction in Belize: An Historical Geography", in *Yearbook Conference of Latin Americanist Geographers* 26 (2000), 103–115; Jennifer L. Anderson, *Mahogany: the costs of luxury in early America* (Cambridge: Massachusetts, Harvard University Press, 2012), 104–124.
132   AGS, SMA Marina, Asientos, leg. 351, Estado de arboladura existente en el embarcadero de Coatzacoalcos.

1782, when a cargo of pine masts and bitumen for the Navy set sail down the Mississippi River in the ship *La Felicidad*.[133]

The Navy began its own felling operations in the region in the summer of 1784, as confirmed by a letter, dated 29 January 1785, which mentions the presence in New Orleans of 280 pieces of masting for the *Real Hacienda*; the operation had been overseen by Naval Lieutenant Juan Antonio Riaño, under the supervision of the intendent and the accountant in the province of Louisiana, Martín Navarro and Juan Ventura Morales, respectively.[134] In the spring of 1786, the New Orleans-based merchants Nicolás Verbois, Santiago Jones, Pedro Bellim, and Esteban Watts proposed an *asiento* to supply the Cuban shipyard with masting in 1787. The proposal evaluated pieces of pine between 69 and 86 cubits long at 180 pesos each; between 30 and 68 cubits long, at 6 reales per foot; and the rest at 4 reales per foot. Francisco Xavier de Morales, fleet commander in Havana, accepted the proposal in October 1786, and felling began in April 1787.[135] Meanwhile, Ignacio Lovio and Juan Baptista Macanti, residents of New Orleans, made an offer to supply ordinary and external planking to the shipyard, and the fleet commander made an interesting observation: "if this planks is as good as that which comes from the North, it would be cheaper for the Spanish shipyards; apart from the fact that the money would not leave H.M.'s dominions; it is worth comparing them in Havana".[136] The offer was, again, accepted, and the contract was in force throughout 1787.

The successful outcome of these early *asientos* encouraged Navy officers in Havana to keep a permanent commissioner in New Orleans to represent this institution in Louisiana. The responsibility fell to Ignacio Lovio, who in the following two years negotiated three contracts. The first was signed with Don Nicolás Verbois for the supply of masting to Havana during 1788, under the following conditions: 167 pesos and 4 reales for each piece exceeding 60 feet in length; and three reales per foot for smaller masts.[137] The second *asiento* was signed with Asahel Levis, resident of Baton Rouge, in April 1788. The contract agreed on the delivery in Havana, within a year, of the following parts: 1,000 boards, 3 inches thick, 12–14 inches wide, and 24–30 feet long; 1,500 external boards, 1 inch thick, 10–12 inches wide, and 24–30 foot long; and 5,000 ordinary

133   AGI, Santo Domingo, leg. 2609, documento núm. 125.
134   AGS, SMA Marina, Asientos, leg. 637.
135   AGS, SMA Marina, Asientos, leg. 637.
136   "si esta tablazón fuese de tan buen servicio como este del Norte, saldría más barata que
       aquella en los arsenales de España; y por descontado no se extraía su importe fuera de los
       dominios de S.M. convendría pues mandar hacer en La Habana alguna prueba de com-
       paración". AGS, SMA Marina, Asientos, leg. 637.
137   Ibid.

boards of the same thickness and width, 12–14 foot long.[138] In contrast with these two short-terms *asientos*, the contract signed by Lorenzo Sigur, from New Orleans, was for a four-year period, beginning in February 1789. The *asentista* committed to supply Havana with masting, planking, and external boarding from Louisiana; pieces that were 78 and a half feet long, or longer, were costed at 160 pesos each, and shorter ones at 3 reales per foot. External boarding, 1.5–3 inches thick, was costed at 6 reales per foot; external boarding was costed at 3 reales per foot, and ordinary planking at 2 reales per foot.[139] The proposals were accepted and yielded good results and kept a steady supply of pine wood for the Havana shipyards.

The success of these contracts prompted the Navy officers to organise a scientific expedition to Louisiana in July 1789. Navy Lieutenant Juan Antonio Riaño's reports mention the following valuable woods:

– *Red and white cypress*: Tree which grows everywhere in upper and lower Louisiana; there are two species, the red and the white [...]. the Red is durable and heavy and sinks in water when freshly cut. The white is very elastic and light but less durable than the other species, and these properties make it appreciated for furniture and masting, also because it has fewer knots. Another species of cypress is known, different from these two. It is durable and close to incorruptible, light and it makes beautiful furniture. These three species are the most beautiful trees in Louisiana [...] they are well proportioned: the white Cypress is generally 4 feet in diameter and between 70 and 110 feet tall; it is the most suitable for the canoes that float on these rivers. These woods are invulnerable on sea and on land, and are often used in civil and naval construction; planking, panelling, and larger pieces, like keeps and staves.[140]

---

138   Ibid.
139   Ibid.
140   "Árbol que crece en todos los parajes de la Luisiana alta y baja hay de dos especies, colorado y blanco [...]. el Colorado dura mucho, es muy pesado, porque se va al fondo del agua cuando esta recientemente cortado. El blanco es muy elástico y ligero, pero de menos duración que la otra especie, es el más estimado para obras de carpintería y arboladura por sus propiedades, como igualmente por ser menos nudoso. Se conoce también otra especie de ciprés, cuya misma no va por ser igual a las dos citadas. Es muy durable y cuasi incorruptible, ligero y propio para muebles que son muy bellos y hermosos. Las tres especies son los más bellos arboles de la Luisiana [...] son bien proporcionadas: el Ciprés blanco es comúnmente de 4 pies de diámetro y de 70 a 110 pies de altura: es el árbol más propio y que más comúnmente se emplea para construir las piraguas y canoas que navegan estos ríos. La madera de estas especies de Árboles es invulnerable en la tierra y en agua, se emplea generalmente tanto en construcciones civiles y navales: tablas, tablero, y piezas más grandes, tajamares y duelas".

– *Pine*: it grows in the upper and the lower Louisiana; it is a first-rate tree, the woods are enormous, and it is of great value for the Navy, but it is very heavy and cannot compare to the cypress, because it does not last as long, so its main use is to produce pitch and tar.[141]
– *Holm oak*: it grows in the highlands of lower Louisiana; it is incorruptible on land and very hard on water; it always grows crooked, but it is very apt for the construction of ships-of-the-line; there are trees so monstrous that a single branch can be the keel for an 80-guns ship.[142]
– *Ash*: it is generally used for the kitchens.[143]
– *Walnut tree*: the black walnut tree is appreciated to make boarding, furniture, and gun carriages.[144]
– *Mulberry tree*: like the European kind, is used in carpentry.[145]
– *Black poplar*: it is unsuitable for carpentry.[146]
– *Cedar*: there are white and red cedars. The wood is nigh incorruptible, and the milk is bitter, so the worms do not attack it. Cedar could be used for naval building and for constructions that are exposed to the sun, water, and air. The branches are thick and it is about 30 or 40 feet tall.[147]
– *Poplar*: it is used to build carts.[148]
– *Sassafras*: the wood is suitable for carpentry, and in naval building it could be used for the superstructure, as it is light, durable, and resistant to exposure.[149]
– *Sugar tree*: good wood for light carpentry, it is used for musket butts.[150]

Despite the great interest of the Navy Department in Havana in Louisiana wood, the business slowly went into decline because of the crisis that overcame the

---

141 "crece en la alta y baja Luisiana, es árbol de primera magnitud, los bosques son inmensos, de gran uso en la armada, pero es muy pesado, y no iguala al Ciprés pues dura muy poco tiempo, por lo que el principal uso es para sacar brea y alquitrán".
142 "crece en los países altos de la Baja Luisiana, es incorruptible en la tierra, y en el agua durísimo; crece siempre torcido pero en muy propio para la construcción de navíos y hay árboles tan monstruosos que una sola rama puede dar principal curva de un navío de ochenta cañones".
143 "generalmente se usa para las cocinas".
144 "el nogal negro muy estimado para hacer tablas y piezas propias para muebles y soporte de armas".
145 "es el mismo que en Europa y se usa para obras de carpintería".
146 "poco apropiado para obras de carpintería".
147 "hay blanco y colorados. La madera es casi incorruptible, su leche amarga lo que impide que los gusanos la ataquen. El cedro podría emplearse en la construcción de barcos y otras obras expuestas al sol, agua y aire. Tiene gruesas ramas y su altura es de 30 a 40 pies".
148 "se usa para obras de carretería".
149 "la madera es muy propia para carpintería de blanco se podrá emplear en la arquitectura naval para las obras muertas por ser ligera y dura mucho expuesta al aire".
150 "buena madera para carpintería blanca, se usa para montarías de fusiles". ANRC, Junta de Fomento de la Isla de Cuba, leg. 179, orden 8211.

Navy beginning in 1795. At that time, some cargoes with planking and masting were still arriving, but in smaller quantities than in the 1780s.

11   Wood Sourcing Projects Elsewhere in the Greater Caribbean

Private and Navy-promoted felling projects reached other American regions, for instance, San Blas, El Realejo, Guayaquil, and Concepción, in the South Sea, as well as the woodland areas of Cartagena de Indias, Santa Marta-río Magdalena and Trinidad-Cumaná, in the Caribbean coast (see Map 11).[151]

The Caribbean was the source of different woods for the Compañía Guipuzcoana de Caracas, which had set up a cacao *asiento* in Venezuela in 1728. These woods, however, amounted to little more than a few mahogany pieces for rudders and anchor stocks, as well as some cedar for planking. In 1754, the Secretary of the Navy began taking an interest in the region's woodland resources and considered the possibility of creating a shipyard on the island of Trinidad, with a wood supply base in Cumaná. This and another project presented in 1768, which merely concerned felling, were ruled out for economic and administrative reasons. However, the idea did not go away; for instance, in early 1773, the Navy sublieutenant Gerónimo Franco was sent to survey woodland in the Orinoco River, the coast of the Gulf of Paría, and the island of Trinidad.[152] A year later, an *asiento* was signed with Don Manuel Blanco to supply the Spanish departments with naval wood, including cedar and hardwoods from the area around Cumaná. This contract, however, met the opposition of the province's governor, who disrupted felling operations and the transport of the wood.[153] Finally, after some political developments, felling and deliveries could be resumed, as shown by a report dated February 1779. The new governor,

---

151   Michael E. Thurman, *The Naval Department of San Blas: New Spain's Bastion for Alta California and Nootka 1767 to 1798* (Glendale: Arthur Clark Co., 1967), 57–60; Lawrence A. Clayton, *Los astilleros de Guayaquil colonial* (Guayaquil: Archivo Histórico del Guayas, 1978); María Laviana, *Guayaquil en el siglo XVIII. Recursos naturales y desarrollo económico* (Seville: EEHA, 1987); De Aranda y Antón, "Las maderas de Indias", 217–248; Carlos Martínez Shaw and Marina Alfonso Mola, "Los astilleros de la América colonial", in Alfredo Castillero Calvo and Allan J. Kuethe (eds.), *Historia general de América Latina*, vol. 3, part 1 (Madrid: Editorial Trotta/Paris: Ediciones Unesco, 2001), 279–304; Valdez Bubnov, *Poder naval y modernización del Estado*, 336, 376–379; Wing, *Roots of Empire*, 198–199; Guadalupe Pinzón Ríos, *Hombres de mar en las costas novohispanas. Trabajos, trabajadores y vida portuaria en el Departamento Marítimo de San Blas (siglo XVIII)* (Mexico: UNAM, 2018), 136–144.

152   Museo Naval de Madrid (MNM), leg. Ms. 126, fs. 244–268.

153   AGS, SMA, Asientos, leg. 352, Asiento de don Manuel Blanco.

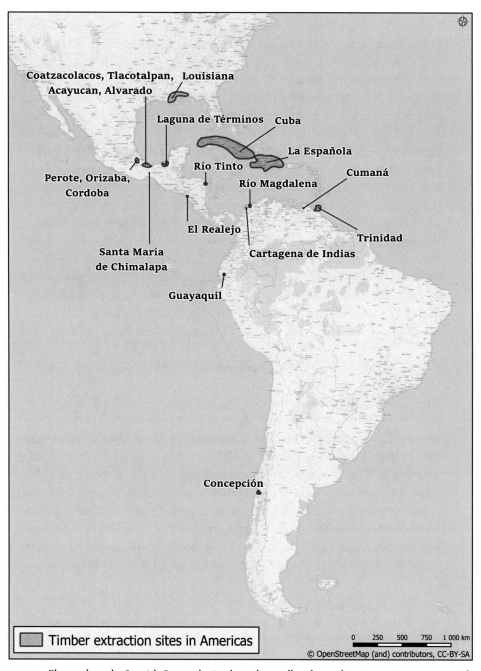

MAP 11   Places where the Spanish Crown obtained wood, as well as those where inspections were carried
out to discover more forests that could provide timber for shipbuilding
SOURCE: LOOK4GIS-LUKASZ BRYLAK BASED ON QGIS SOFTWARE

Máximo du Bouchet, reports the accumulation of dressed pieces of bullytree and nazareno for ships-of-the-line and frigates, pointing out that, after three months in the water, some pieces were beginning to show shipworm; the pieces buried in the sand, on the other hand, were in sound conditions. In April 1779, the king's ship *La Especiosa* was loaded with 360 parts, 181 for ships-of-the-line and 179 for frigates, to an estimated total of 1,841 cubic cubits. The report describes the width, thickness, length, and cubits of each piece, as well as the species out of which they had been prepared. Back of rudders, keels, *galimas* and planking were of nazareno, pardillo, maría, cedar, male cedar, guatamare, charo, paneque, pink poui, macao, oak, cacao cimarrón, caro, aco, araguaney, laurel, chupon, carob tree, and cabima. Finally, after loading all the pieces, the ship left for Cádiz on 9 April 1779.[154] Wood was still being sourced in the region in the 1780s, but less intensively, owing to transport issues; a number of pieces were lost for lack of shipping. In 1788, José María Chacón, governor of Trinidad, organised an expedition to the forests of the Orinoco, where they found wood suitable for naval construction, but its extraction was not undertaken. Finally, logistic problems and the high supply costs for the Navy of the felling of the woodlands of Trinidad, the Gulf of Paría, and Cumaná concluded this exploitation in the 1790s.[155]

12   Woodland Areas in Cartagena de Indias and the Magdalena River

Since it became the winter harbour for the Tierra Firme fleet in 1561, Cartagena de Indias turned into an important enclave for the repair of the ships running the *Carrera de Indias*. Generally, the wood for these repairs was sourced in the forests near the harbour.[156] In the 18th century, Cartagena also became an important base for the Navy and the coastguards. From the 1760s to the 1780s, the main naval base in Tierra Firme was financially dependent on the department in Havana.[157] Around this same period, the first ideas to source wood for the Spanish arsenals in the region emerged. The first firm proposal was put forward by Don Juan Agustín Pardo, resident of Cartagena de Indias, in 1767; Pardo offered "all manner of curved and straight parts from the city and

---

154   Ibid., leg. 362, Madera de Cumaná.
155   Archivo General de Marina "Álvaro de Bazán" (AGMAB), Expediciones a Indias, leg. 07.113.
156   Haring, *Trade and Navigation*, 201–203; Enrique Marco Dorta, *Cartagena de Indias: puerto y plaza fuerte* (Bogotá: Fondo Cultural Cafetero, 1988), 34–36; Martínez Shaw and Alfonso Mola, "Los astilleros de la América colonial", 290–292; Sergio Paolo Solano de las Aguas, "El Apostadero de la marina de Cartagena de Indias, sus trabajadores y la crisis política de la Independencia", *Economía & Región* 9, no. 1 (2015), 209–243.
157   Valdez Bubnov, *Poder naval y modernización del Estado*, 393–397.

its surroundings".[158] He pointed out that the cedar from Cartagena, the most durable in America, could be of great value to build or repair ships-of-the-line and other vessels. He also committed to deliver the wood in Cádiz, for which he requested permission to buy a Swedish or Dutch fluyt.[159]

Although the proposal was not accepted, it opened the door to future projects and contracts, for instance, one signed by the *Real Hacienda* and Don Antonio Segundo Mozin on 10 February 1776, for 2,000 pieces for construction, amounting to a total of approximately 30,000 cubic cubits, at 20 reales per cubit. The contract was for three years and was signed under the guarantee of Don Manuel Josef de Vega, the factor of the Compañía de Negros y de Comercio de Cádiz. The contractor and his partners committed to deliver no fewer than 700 prepard pieces per year. The felling took place on the banks of the Magdalena River, whence the rafts were sent to Cartagena de Indias via Santa Marta. The pieces had to be pitch perfect, without cracks or rot; the knees of felled trees were to be used to the utmost, and the wood was to be stored on the beach of Tinglado. Interestingly, the contract ruled out any down payments; all cash was to follow the delivery of the annual part quota. In contrast, the contract allowed the contractor and the *Real Hacienda* to sign other contracts outside the conditions agreed between the parts. An interesting clause bound the contractor to place "a white flag with the king's arms in the felling area, to distinguish it from the areas where private agents operate".[160] This highlights the relevance of the agreement and the prestige attached to works undertaken on behalf of the king.

In autumn 1775, Manuel Josef de Vega sent three rafts from which the Navy officers selected 112 pieces of mahogany and cedar, which were stored under the shed. Fernando de Lortia, Navy brigadier and commander of the coast-guard of Tierra Firme, wrote to Julián de Arriaga, Minister for the Navy, to say that he was waiting for two ships that had left Cádiz to ship troops to Puerto Rico, which would touch in Cartagena on the return leg to collect the wood and take it to La Carraca. He reported that, by late May 1776, he should have more or less 800 pieces ready for Spain, while indicating that a bigger warehouse was needed.[161] It is also worth pointing out that, during their patrols up and down the coast of New Granada in 1775, coastguard captains Francisco de Luna and

---

158 "todo género de maderas de construcción de vuelta y derecha en dicha ciudad y sus inmediaciones".

159 AGS, SMA, Asientos, leg. 624, Madera de Cartagena de Indias.

160 "poner en los sitios donde establezca la corta, la bandera blanca con el escudo de armas del Rey para distinguirlo de los lugares en que cortan los particulares". Ibid., Asiento de don Antonio Segundo Mozin.

161 Ibid., El comandante de guardacostas de Tierra Firme.

Joaquín de Cañaveral had compiled information about the woodland between Portobelo and Santa Marta, and especially the jungle regions of the Darién, around Cartagena de Indias, and the Magdalena River.[162]

According to the information provided by Captain Fidel de Eslava, sourcing wood in these regions was difficult, because access to good trees was hard and private felling mostly targeted the largest trees; in the Darién, this was compounded by the hostility of "barbarous Indians".[163] For these reasons, the merchants of New Granada were not particularly interested in the business. In the following years, Mozin's contract kept running its course and new felling operations began in the regions of Sinú and Lorica, especially parts for repairs in Cartagena de Indias. Finally in 1786, Juan Agustín proposed a four-year contract to supply 60,000 cubic cubits of dressed and curved pieces of cedar and hardwood.[164]

---

162   AGS, SMA, Asientos, leg. 352, El comandante de guardacostas don Fidel de Eslava.
163   "indios bárbaros".
164   AGMAB, Arsenales, Maderas, leg. 3774.

General Conclusions

The second half of the 18th century, especially the reign of Charles III (1759–1788), is regarded as the period of the greatest political – economic – scientific – military and naval splendour of the Spanish Crown. This distinguished king, with the support of his enlightened ministers and royal officials, undertook numerous projects to develop his empire, both in the Iberian peninsula and his overseas possessions. The modernization of the navy, with the target of turning it into a worthy rival of the British Royal Navy, which at the time dominated the seas and threatened the political – commercial interests of the Bourbon state, played a central role in this programme. An additional contributor to this ambitious project of naval renovation was the Spanish defeat in the Seven Years' War (1756–1763), during which Spain temporarily lost control of Manila and Havana, two strategic ports for the navigation system that linked the Asian, New Spanish, Caribbean, and Iberian markets. These losses forced the Spanish to hand over the province of Florida to the British and demonstrated the operational shortcomings of the American armies and the squadron deployed to protect the Cuban capital. In 1764, administrative (with the creation of the *Intendencia*), military (creation of the professionalization of the militia), and naval (change from the British to the French ship construction system) reform programmes were implemented in Cuba.

This reformist enterprise also affected the navy more broadly, beginning with a large-scale shipbuilding programme, especially 74- and 112-gun ships-of-the-line. Between 1765 and 1794 these efforts were overseen by two enlightened shipbuilders, Francisco Gautier (1765–1782) and José Joaquín Romero y Fernández de Landa (1782–1794), who respectively held the posts of General Engineer and Director of the Navy Engineer Corps, which had been created in 1770. During their time in office, 21 74-gun ships and 10 112-gun triple-decker ships were built in El Ferrol (12 ships), La Habana (7 ships), Guarnizo (6 ships), and Cartagena (6 ships). In addition, the royal shipyards also launched two 80-gun ships, three 64-gun ships, and 29 frigates armed with 24 to 38 pieces of ordnance.[1] This enormous output was possible because of the administrative and financial reorganization of the navy undertaken during Charles III's reign.

---

1 José María Blanco Núñez, *La Armada Española en la segunda mitad del siglo XVIII* (Madrid: IZAR Construcciones Navales, 2004); Enrique García-Torralba Pérez, *Las fragatas a vela de la Armada Española 1600–1850. Su evolución técnica* (Madrid: 2011); José María Sánchez Carrión, *De constructores a ingenieros de marina. Salto tecnológico y profesional impulsado por Francisco Gautier* (Madrid: Fondo Editorial de Ingeniería Naval, 2013).

© KONINKLIJKE BRILL BV, LEIDEN, 2024 | DOI:10.1163/9789004689640_007

The impetus was still there in the early years of Charles IV's reign, but 1794 may be regarded as the closing date of this period of apogee for the Spanish Navy; afterwards, Spanish naval construction entered a period of logistic, administrative, and financial crisis that finally drove Spain's naval golden age to an end.

However, during these 29 years (1765–1794), especially in the 1770s and 1780s, the Spanish Navy peaked under the direction of secretaries Julián de Arriaga y Ribera (1754–1776), Pedro González de Castejón (1776–1783), and Antonio Valdés y Bazán (1783–1795), becoming the second most powerful navy in the world, after the British Royal Navy. This was possible thanks to the administrative skill of ministers, intendents, and royal officials, but also to the state's expense priorities. This gave the Bourbon officials leverage to negotiate with national and foreign merchants and guarantee the supply of wood from Spain, Europe, and America. By 1794, the construction programme had yielded 36 ships-of-the-line and 29 frigates of various sizes.

It is important to remember that this success was possible because the Spanish monarchy in the Modern Age implemented a system for naval supplies based on contractual relationships. This administration was the answer to the growing maritime needs of the Habsburgs' and the Bourbons' war policies. Naval, but also the military, resources were directed through contracts, which also protected the Spanish laws. With this policy, the crown became a "contractor state", which resorted to various legal contracts for naval supplies and services with individuals and other establishments networks. From the moment the crown contracted for its needs, it was also obliged to respect and enforce the laws that regulated its *asientos*. Through those agreements, the Hispanic monarchy determined the legal framework and thereby made the activity predictable for the crown and the contractors. The normalization of the mobilization of naval and military resources with contracts reduced the uncertainty of the model based on the feudal and family relationships of the Middle Ages. The regulation of naval supply contracts by the bureaucratic apparatus allowed the extension of Spanish Crown control on the goods and business partners' collaboration. This is an important issue because, throughout the Modern Age, it allowed the monarchy to establish the limits of its authority so that the contract could be national or imperial. On several occasions, the geographical location and limitation of actions of the mobilization of naval resources could be determined by contracts.[2] In the second half of the 18th century, it was clearly demonstrated by the agreements signed with different national and international companies to transport and commerce timbers

---

2   Rafael Torres Sánchez, *Military Entrepreneurs and the Spanish Contractor State*, 13–40.

from the southern Baltic and with Creole elites from the Viceroyalty of New Spain to log wood.[3]

Spain's imperial challenges primarily revolved around those regions outside the Iberian peninsula, with which Spain had close political, administrative, and commercial links to obtain a supply of timber. The southern Baltic and the colonies in the Greater Caribbean territories had been connected by sea with Spain for centuries, but it was not until the 18th century that Spain realized their potential in terms of a large-scale timber supply policy. In the case of the southern Baltic area, *asientos* were granted to national and more often foreign merchants as shown by examples of contracts signed with Gil de Meester (Dutch merchant house), Simón de Aragorri, Miguel de Soto (merchant of Irish origin), Felipe Chone, Carlos María Marraci, Pedro Normande and Herman & Ellermann & Schlieper (German merchants), Rey and Brandenburg (Swedish merchants), and Gahn (Swedish merchant). These businessmen had extensive influence in the ports of the Baltic and North Sea and guaranteed the supply of wood needed to build warships in the royal shipyards in El Ferrol, Cádiz-La Carraca, and Cartagena. In addition, they established contacts with local merchants. Good examples of this cooperation are with Thomas and Adrián Hope (based in Amsterdam), Antonio de Cuyper (Dutch commercial representative in Gdańsk/Danzig), Ignacio Jacinto Mathy (French consul in Gdańsk/Danzig), Juan Felipe Schultz (merchant in Gdańsk/Danzig), Herman Fromhold (merchant in Riga), and Blankenhagen Oom & Co (commercial house in Riga). Timber purchases in the Baltic region were also supported by Spanish diplomats, such as Juan Manuel de Uriondo (consul in Amsterdam), the Count of Aranda (ambassador in Poland), Luis Perrot (consul in Gdańsk/Danzig), and Antonio Colombí y Payet (consul in Saint Petersburg) they explored the markets potential of Baltic ports and convinced local merchants to trade timber with Spain.

Those networks that Spanish contractors formed in the second half of the 18th century in the southern Baltic area opened the tap of pine, fir, and oak timber for masting, planking, and beams, from ports of Szczecin/Stettin, Gdańsk/ Danzig, Königsberg, Memel, Riga, and Saint Petersburg. The harbours connected through navigable rivers like the Oder, Vistula, Pregoła, Łyna/Alle, Neman, and Daugava with abundant virgin forest masses in their hinterland in kingdoms of Prussia, the Polish-Lithuanian Commonwealth, and Russia Empire. There, forest owners or tenants supplied all sorts of timber, as attested by the invoices found in the archives. In the case of the Polish-Lithuanian Commonwealth,

---

3 It must be remembered that another important source from which timber for the shipbuilding was imported is the Mediterranean. Due to various limitations, it was not featured in this book.

the local magnate families like the Radziwiłł, Massalski, Swadkowski, and Wyszynski, and their represents like Dernattowitz, and Reuhutz (Jewish merchants) were closely connected with the timber provision policies of the main European naval powers, including Spain. Those crowns were fairly dependent on the timber supplies from the territories of the southern Baltic. In the case of Spain, as exemplified by contracts by Gil de Meester, Simón de Aragorri, Miguel de Soto-Felipe Chone, Carlos María Marraci, Pedro Normande, and Gahn show, the *asiento* system operated efficiently for the Marina Real's needs and kept a flow of the Baltic timber to preserve the naval departments of El Ferrol, Cádiz-La Carraca, and Cartagena working at appropriate capacity.

Another policy used by the Bourbon state to get timber was by logging under the direct control of the crown in their American colonies. During the 18th century, Havana was the epicentre of Spanish naval tradition in America and became a major hub for shipbuilding and repair. However, it did not acquire a leading role for the Royal Navy until the earliest shipbuilding *asientos* granted to Manuel López Pintado (1713–1717), Juan de Acosta (1717–1740), the Real Compañía de La Habana (1740–1750), and finally the Intendancy of Navy, which managed ship construction during the second half of the 18th century. Despite this, American wood resources were underexploited, as attested by the essays written by several economists and politicians, such as Jerónimo de Uztáriz and José del Campillo y Cossío, who tried to promote naval construction in other regions of the Greater Caribbean, such as Coatzacoalcos, Campeche, and Cartagena de Indias, while emphasising the availability of major cedar, mahogany, and sabicú forests in their vicinity. In the event, these locations only witnessed minor operations (e.g. the attempt to build ships-of-the-line in Coatzacoalcos was a total failure and a waste of economic resources), ship repairs, and wood extraction for the Havana shipyards, such as cutting seasons of pine masting in Chimalapas.

As in the case of the southern Baltic region, a period of greater activity, surveying and harvesting timbers in the Viceroyalty of New Spain and the Caribbean area began in the 1770s, when viceroys Antonio María de Bucareli y Ursúa (1771–1779) and Martín de Mayorga (1779–1783) sponsored the various expeditions of military engineer Miguel del Corral and Royal Navy officers Joaquín de Aranda, Luis Fernández, and Miguel Sapiain to the forests and rivers in the provinces of Veracruz, Oaxaca, Tabasco, Laguna de Términos, and Campeche to assess the availability of timber for naval construction. The result of these expeditions was the several *asientos* granted to the local elites in the 1780s: José Jiménez, militia captain in Tuxpan; Esteban Bejarano, merchant from Veracruz; Pedro Moscoso, vecino of Acayucan; Ramón Carvallo, vecino of Tlacotalpan; Francisco Sánchez de Burgos, subdelegate of La Antigua; Joseph

Nicolás Sánchez, river carpenter in presidio del Carmen, through the viceroy-alty administration, they signed logging contracts with the Spanish Crown. The largest *asientos* were awarded between 1784 and 1787 when the members of the Veracruz elite committed to supplying approximately 30,000 cubic cubits of carved parts in cedar, mahogany, and sabicú. For this, the local contractors received 78,000 pesos de a ocho reales. However, storage difficulties at the mouth of the Alvarado and Coatzacoalcos rivers and the beaches of Veracruz, and constant shipping deficiencies that hampered the transport of the timbers from Veracruz to the naval departments in Spain and Cuba meant that only 11,745 cubits and 229 parts could be delivered to El Ferrol, Cádiz-La Carraca, and Havana, while 5,193 cubits and 198 parts were left lying in the warehouses of Veracruz. Owing to these issues, the contracts were cancelled in 1787 and large-scale harvesting of wood in New Spain came to an end. These analysed examples show that the Spanish Crown, without strong support from merchant ships, was unable to pursue an effective policy of obtaining timber from its overseas possessions in America. Comparing the contracts from the southern Baltic with those from the Viceroyalty of New Spain, it is evident that the strength of the first region was in the effective organization of transport between the Baltic ports and Spain's royal shipyards.

It is worth emphasising that the exploration of American forests by royal officers was also carried out in other regions, where attempts were also made to establish logging under royal patronage or by awarding contracts to American Creoles. For this reason, some logging campaigns were undertaken in other regions of the Greater Caribbean. Those commenced in Cumaná by the Real Compañía de Caracas bore fruit, and the timber harvested was used for the company's own needs. Another contract that resulted in a substantial output was one granted to Juan Agustín Pardo to obtain 60,000 cubits of cedar between 1786 and 1790. The contractor delivered nearly one-third of this volume, but the *asiento* was cancelled because the Royal Navy could not provide solid transport for subsequent lots of wood. Finally, a venture that met the crown's expectations was contracts awarded to Louisiana merchants to cut and deliver timber from that province to the shipyards in Havana. This state of affairs was demonstrated by contracts signed in the 1780s with Nicolás Verbois, Santiago Jones, Pedro Belli, Esteban Watts, Ignacio Lovio, Juan Baptista Macanti, and Asahel Levis, merchants from New Orleans and Baton Rouge. Of this group, Verbois and Lovio proved to be effective suppliers of pine and cedar wood. Analysing these contracts, it is obvious that these businessmen were not of Spanish origin, which indicates that in Louisiana the Spanish Crown traded with foreigners who settled in these places, ceded to Spain under the Treaty of Paris in 1763. A similar situation existed in the southern Baltic region, where the main shareholders in the timber trade were the Dutch and Scandinavians. This is another indication of the weakness of the Spanish merchant fleet, which did not provide sufficient support in the trade of important strategic materials such as wood.

Relating to timber supply policies in the Southern Baltic and the Viceroyalty of New Spain, it seems clear that although the Bourbon state had the financial funds to create a navy capable of protecting the maritime routes with the American colonies and could use the effective contract system linked with well-functioning merchants—the shortcomings of the Spanish merchant marine fatally undermined its efforts and the state proved unable to provide the Royal Navy's shipping needs. In the case of the southern Baltic, this was compensated for by the wide use of freight on ships from Holland, Great Britain, Denmark, and Sweden, among other countries. Ironically, the timber was very often hauled by the British, who were the enemies of the Spanish Crown. In the American case, the option of using a network of foreign merchants was not available, as this would contravene Spain's monopolistic policies that limited transatlantic trade to the Spanish market. Not even the use of the king's cargo vessels proved sufficient or profitable enough to guarantee a regular timber supply. For these reasons, and despite the improved knowledge of American woodland—where excellent forests of cedar, mahogany, and other hardwoods, were identified—American woods were of limited use for naval construction. Interestingly, the British Royal Navy managed to use forest resources in North America, where before the outbreak of the Revolutionary War in 1775, one-third of British merchant ships, were built in the colonies. Oak and pine timber were also sent in huge quantities to shipyards located in England.[4] Similarly, the Portuguese colony in Brazil shipped large quantities of tropical timber from provinces such as Rio de Janeiro, Bahia, Alagoas Pernambuco, and Paraíba to the Lisbon shipyard.[5] These examples clearly show that it was possible to successfully use American wood in European shipbuilding, but not on a large scale in the Spanish case due to the meagre merchant fleet.

In addition to the economic and commercial aspects, another issue that the book analyses is the management of forest resources so that quality timber could be harvested. In the case of the southern Baltic, these matters rested in the hands of royal officials, as shown by the cases from the kingdoms of Prussia and Poland, or were managed by forest owners, as occurred in the

---

4   David Kirby and Merja-Liisa Hinkkanen, *The Baltic and the North Seas* (London: Routledge, 2000), 97.

5   Shawn W. Miller, *Fruitless Trees: Portuguese Conservation and Brazil's Colonial Timber* (Stanford: Stanford University Press, 2000), 183–208.

Polish-Lithuanian Commonwealth, where some magnate families such as Zamoyski and Radziwiłł conducted their own forestry policies. The first family pursued a sustainable logging and afforestation policy, while the Radziwiłłs were not interested in protecting their forests because they saw only economic benefits from them. In the case of Spanish America, where the law given by Spain was in practice the forest reform (*Ordenanza*) of 1748—supervised by the Royal Navy—was not introduced, but its modified version for all territorial divisions, such as *corregimiento*, published in December of the same year. By analysing the historical documentation concerning the inspections of mountains and timber logging in the Greater Caribbean region, it was observed that the royal officials in charge of these missions were aware of the protection and sustainable use of the forests, not for the purposes of deforestation these areas but in order to ensure timber reserves for the shipbuilding industry of the *Marina Real*.

Finally, I would like to emphasise that this book presents an analysis of the political, economic, and military efforts of the Bourbon state to significantly reinforce its naval power during the second half of the 18th century. Looking at the Spaniards' determinations to establish solid networks of timber provision—from outside of the Iberian peninsula—it must be said that the *Marina Real* was dependent on forest resources and other naval provisions from remote, peripheral regions—such as those presented in the book—which were located in the hinterlands of the Southern Baltic, the Viceroyalty of New Spain, and the Greater Caribbean. For this reason, it is important to review the globalization process for Early Modern History and to begin to write it from a new angle, where the local perspective must be perceived as a motor for global imperial connections.

Appendix

I would like to present some drawings depicting wood, woodworking, and its use in shipbuilding, which were created over many decades by Juan José Navarro y Búfalo, Marquis of Victoria, who was a Spanish military officer who served as the first Captain General of the Navy from 1750 to 1772.

These drawings are taken from his greatest work produced between 1719 and 1756, the *Diccionario demostrativo de la arquitectura naval antigua y moderna* as it is generally known, is a veritable encyclopaedia of everything to do with ships. The Dictionary is divided into four main sections, relating to shipbuilding, rigging, and ship equipment, understood in its broadest sense, and a part devoted to various special features.

The part on shipbuilding includes aspects related to the procurement of timber, from its location, selection, felling, and seasoning, to its arrangement to form the structure of an English-style ship. All aspects of rigging are covered, from the preparation and rigging of masts and spars, through the arrangement of standing and running rigging, concerning types of rigging, to the cutting and handling of sails. A third block could include the equipment of the ship, including the launches and *falúas*, the tools of the carpenters and other trades of the ship and the shipyard, the materials used for the construction, ironwork, nailing, and everything related to the equipment of the ship. Of particular importance in this section is the study of the artillery and objects relating to the ship's trades, a veritable catalogue of life on board. The dictionary is completed by a good number of particularities and curiosities that can be grouped in a fourth block, which begins with the delineation of the ancient monuments of the ships that were used and have been found, in marbles, obelisks, paintings, and medals, to which can be added a good number of plates relating to the layout of arsenals, machines, cabins, and infrastructures.

Currently, this shipbuilding album is located in the Naval Museum of the Spanish Navy in Madrid.

© KONINKLIJKE BRILL BV, LEIDEN, 2024 | DOI:10.1163/9789004689640_008

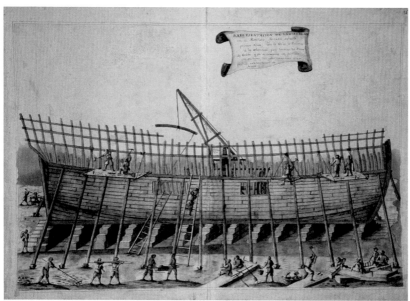

FIGURE 21   Representation of a frigate in the shipyard
SOURCE: SPAIN, MINISTRY OF DEFENCE. ARCHIVO HISTÓRICO DE
LA ARMADA SEDE "ELCANO". *DICCIONARIO DEMOSTRATIVO CON LA
CONFIGURACIÓN* (SHEET 15)

FIGURE 22   English method of cutting timber using large saws
SOURCE: SPAIN, MINISTRY OF DEFENCE. ARCHIVO HISTÓRICO DE
LA ARMADA SEDE "ELCANO". *DICCIONARIO DEMOSTRATIVO CON LA
CONFIGURACIÓN* (SHEET 28)

FIGURE 23   Different figures of tree trunks and branches to give an idea of how to appropriate them to all the parts that go into the construction of a ship
SOURCE: SPAIN, MINISTRY OF DEFENCE. ARCHIVO HISTÓRICO DE LA ARMADA SEDE "ELCANO". *DICCIONARIO DEMOSTRATIVO CON LA CONFIGURACIÓN* (SHEET 21)

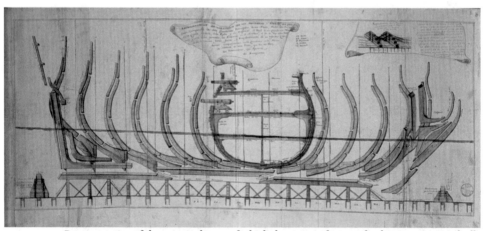

FIGURE 24   Representation of the principal parts of which the 12 main frames of a ship are composed, all the parts of the keel, stern and stem, and the members of the main frame
SOURCE: SPAIN, MINISTRY OF DEFENCE. ARCHIVO HISTÓRICO DE LA ARMADA SEDE "ELCANO". *DICCIONARIO DEMOSTRATIVO CON LA CONFIGURACIÓN* (SHEET 8)

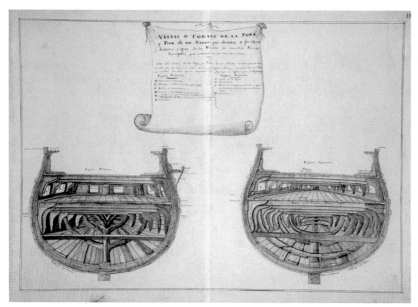

FIGURE 25   View of a stern and bow of a ship from the inside, showing most of its parts
SOURCE: SPAIN, MINISTRY OF DEFENCE. ARCHIVO HISTÓRICO DE
LA ARMADA SEDE "ELCANO". *DICCIONARIO DEMOSTRATIVO CON LA
CONFIGURACIÓN* (SHEET 13)

FIGURE 26   Preservation of construction timbers in stacks, and sheds; recognition after
felling, of their quality and manner of placing them anywhere so that they
were ventilated
SOURCE: SPAIN, MINISTRY OF DEFENCE. ARCHIVO HISTÓRICO DE
LA ARMADA SEDE "ELCANO". *DICCIONARIO DEMOSTRATIVO CON LA
CONFIGURACIÓN* (SHEET 26)

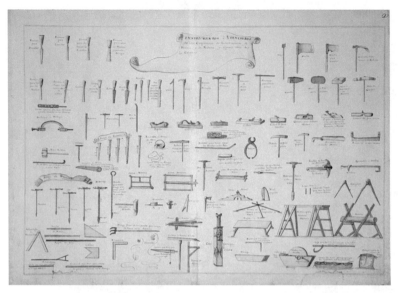

FIGURE 27   Tools and utensils used by carpenters in construction and
shipbuilding, such as a set of drill, hammers, screwdrivers, chisels, etc.
SOURCE: SPAIN, MINISTRY OF DEFENCE. ARCHIVO HISTÓRICO
DE LA ARMADA SEDE "ELCANO". *DICCIONARIO DEMOSTRATIVO
CON LA CONFIGURACIÓN* (SHEET 27)

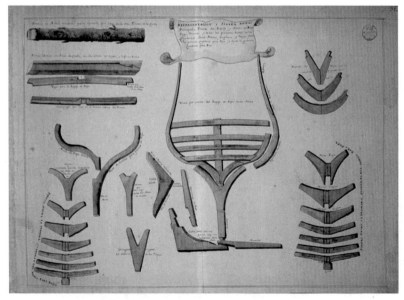

FIGURE 28   Representation and figure of the main parts of a ship's transom, fins
and plans, bowsprits, curved *orcazes*, sleepers and hogsheads, *orcón*
and coral curve
SOURCE: SPAIN, MINISTRY OF DEFENCE. ARCHIVO HISTÓRICO DE
LA ARMADA SEDE "ELCANO". *DICCIONARIO DEMOSTRATIVO CON LA
CONFIGURACIÓN* (SHEET 36)

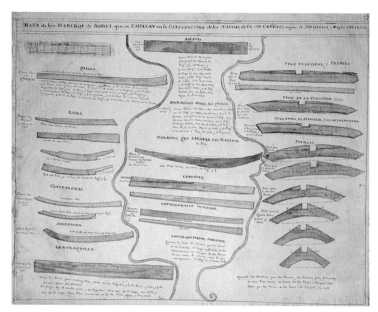

FIGURE 29   Pieces of oak and elm wood, which are used in a ship of 68 or 70 guns
of English system being the same as those of the frigates, although in
smaller proportion
SOURCE: SPAIN, MINISTRY OF DEFENCE. ARCHIVO HISTÓRICO
DE LA ARMADA SEDE "ELCANO". *DICCIONARIO DEMOSTRATIVO
CON LA CONFIGURACIÓN* (SHEET 127)

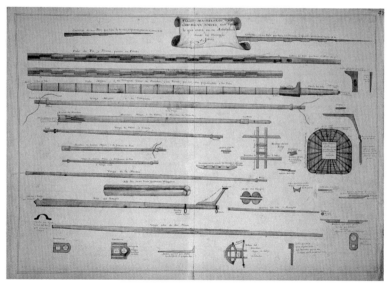

FIGURE 30   Representation and figure of the formation of the masting of a ship
such as masts, spars, spars, bends, bowsprits, etc.
SOURCE: SPAIN, MINISTRY OF DEFENCE. ARCHIVO HISTÓRICO
DE LA ARMADA SEDE "ELCANO". *DICCIONARIO DEMOSTRATIVO
CON LA CONFIGURACIÓN* (SHEET 42)

Glossary of Spanish Words

Alcabala Taxation of the percentage of the price paid to the tax authorities by the seller in the contract of sale

Alcaldía Mayoralty

Alcaldía Mayor Mayor's Office

Almirantazgo Admiralty

Asiento contract

Asentista contractor

Arbolillos baulks

Baos beams

Berlingas spars

Busardas Breast-hook

Caja real place where gold, silver, precious stones, money and other goods were deposited and disposed of during Spanish domination in the Americas

Caminantes walkers

Carrera de Indias Spanish navigation system between the Metropolis and the Americas

Cinco Gremios Mayores pre-capitalist institution, in a way that was compatible with the mercantilist economic policy of the Spain ancient regime

Corregidor Mayor

Contra aletas Fashion-piece

Curvas knees

Gambotas Gaunter timber or head timber

Genol futtock

Gimelga fish

Granos small silver coin

Hacienda farm

Hacendado owner of the farm

Indio Indian

Lechuguilla plant species of Sierra Madre forests

Maravedís de vellón copper coin

Marina Real Royal Navy

Mestizo Mestizo

Ordenanza Ordinance

Pardo mulatto or person of colour

Pesos, pesos de a ocho real silver coin

Relación relation

Reales de plata silver coin

Reglamento de sueldos   Salary Regulation

Real Hacienda   Royal Treasury

Real de vellón   copper coin

Realengo   real property

Recopilación de Indias   set of laws established by the Spanish crown for the West
    Indies

Ramajeros   cleaners

Recogedores   pickers

Tabloncillo   small board

Tinglado   solid wooden or masonry structure where timbers and carved pieces were
    stored

Tomín   small silver coin

Tejamanil   roof tile

Tozería   set of chunks

Tozas   chunks

Vara   unit of length used by the Spanish

Vecino   neighbour

Bibliography

Albion, Robert G. *Forests and Seapower*. Cambridge: Harvard University Press, 1926.

Alcántara López, Álvaro. "Redes sociales, prácticas de poder y recomposición familiar en la provincia de Acayucan, 1784–1802." In *Redes sociales e instituciones comerciales en el imperio español, siglos XVII a XIX* edited by Antonio Ibarra and Guillermina del Valle Pavón, 215–248. México: Instituto Mora, 2007.

Anderson, Jennifer L. *Mahogany: the costs of luxury in early America*. Cambridge, MA: Harvard University Press, 2012.

Andrade Muñoz, Germán Luis. *Un mar de intereses, la producción de pertrechos navales en Nueva España, siglo XVIII*. México: Instituto Mora, 2006.

Anusik, Zbigniew and Andrzej Stroynowski. "Radziwiłłowie w epoce saskiej. Zarys dziejów politycznych i majątkowych." *Acta Universitatis Lodziensis* 33 (1989): 29–58.

Aragón Ruano, Álvaro. *El bosque guipuzcoano en la Edad Moderna: aprovechamiento, ordenamiento legal y conflictividad*. Donostia: Sociedad de Ciencias Aranzadi, 2001.

Aragón Ruano, Álvaro. "Un choque de jurisdicciones. Fueros y política forestal en el Pirineo occidental durante el siglo XVIII." *Obradoiro de Historia Moderna* no. 28 (2019): 135–162.

Aragón Ruano, Álvaro. "Siete siglos de sostenibilidad forestal en Guipúzcoa (siglos XIII–XIX)." *Manuscrits. Revista d'Història Moderna* 42 (2020): 65–88.

Aragón Ruano, Álvaro. "Soberanía y defensa de la riqueza forestal en la frontera vasconavarra con Francia durante el siglo XVIII." *Memoria y Civilización* 25 (2022): 423–450.

Aranda y Antón, Gaspar de. *Los bosques flotantes. Historia de un roble del siglo XVIII*. Madrid: ICONA, 1990.

Aranda y Antón, Gaspar de. "Las maderas de Indias.", *Asclepio* 45, no 1 (1993): 217–248.

Aranda y Antón, Gaspar de. *El camino del hacha: la selvicultura, industria y sociedad: visión histórica*. Madrid: Ministerio de Medio Ambiente, 1999.

Aranda y Antón, Gaspar de. *La carpintería y la industria naval en el siglo XVIII*. Madrid: Instituto de Historia y Cultura naval, 1999.

Artíñano y Galdácano, Gervasio de. *La arquitectura naval española (en madera)*. Madrid: Oliva de Vilanova, 1920.

Asprey, Robert B. *Frederick the Great: The Magnificent Enigma*. New York: Ticknor & Fields, 1986.

Åström, Sven-Erik. "English Timber Imports from Northern Europe in the Eighteenth Century." *Scandinavian Economic History Review* 18 (1970): 12–32.

Åström, Sven-Erik. *From Tar to Timber: Studies in Northeast European Forest Exploitation and Foreign Trade 1600–1860*. Helsinki: Societas Scientiarum Fennica, 1988.

Attman, Artur. *The Russian and Polish Markets in International Trade, 1500–1650*. Gothenburg: Institute of Economic History of Gothenburg University, 1973.

Attman, Artur. *The Bullion Flow between Europe and the East, 1000–1750*. Gothenburg: Kungl. vetenskaps-och vitterhets-samhället, 1981.

Aymes, Jean René. *La guerra de España contra la Revolución Francesa (1793–1795)*. Alicante: Instituto de Cultura Juan Gil-Albert-Diputación de Alicante, 1991.

Bamford, Paul W. "France and the American Market in Naval Timber and Masts, 1776–1786." *Journal of Economic History* 12, no. 1 (1952): 21–34.

Bamford, Paul W. *Forest and French Sea Power. 1660–1789*. Toronto: University of Toronto Press, 1956.

Bankoff, Greg. "The Tree as the Enemy of Man: Changing Attitudes to the Forests of the Philippines, 1565–1898." *Philippine Studies* 52, no. 3 (2004): 320–344.

Bankoff, Greg and Peter Boomgaard (eds.). *A History of Natural Resources in Asia: The Wealth of Nature*. Basingstoke: Palgrave Macmillan, 2007.

Bao, Maohong. "Environmental History and World History." *Journal of Regional History* 2, no. 1 (2018): 6–17.

Barbier, Edward B. *Natural Resources and Economic Development*. Cambridge: Cambridge University Press, 2005.

Barbier, Jacques A. "Indies Revenues and Naval Spending: The Cost of Colonialism for the Spanish Bourbons, 1763–1805." *Jahrbuch für Geschichte Lateinamerikas* 21, no. 1 (1984): 171–188.

Baudot Monroy, María. *La defensa del Imperio. Julián de Arriaga en la Armada (1700–1754)*. Madrid: Ministerio de Defensa-Universidad de Murcia, 2013.

Béthencourt Massieu, Antonio. "Arboladuras de Santa María de Chimalapa Tehuantepec en las construcciones navales indianas 1730–1750." *Revista de Indias* 20 (1960): 65–101.

Biernat, Czesław *Statystyka obrotu towarowego Gdańska w latach 1651–1815*. Warsaw: PWN, 1962.

Binerowski, Zbigniew. "Transport wiślany w dawnej Rzeczypospolitej." In *Dolina Dolnej Wisły*, 283–297. Wrocław: Wydawnictwo Polskiej Akademii Nauk Ossolineum, 1982.

Blanco Núñez, José María. *La Armada Española en la segunda mitad del siglo XVIII*. Barcelona: IZAR Construcciones Navales, 2004.

Bogucka, Maria. "Żegluga bałtycka w XVII–XVIII wieku w świetle materiałów z archiwum w Amsterdamie." *Zapiski Historyczne* 82, no. 4 (2017): 123–137.

Bogucka, Maria. *Gdańsk jako ośrodek produkcyjny w XIV–XVII wieku*. Warsaw: PWN, 1962.

Bogucka, Maria. *Handel zagraniczny Gdańska w pierwszej połowie XVII wieku*. Warsaw: Ossolineum-PAN, 1970.

Bogucka, Maria and Henryk Samsonowicz. *Dzieje miast i mieszczaństwa w Polsce przedrozbiorowej*. Wrocław: Ossolineum, 1986.

Boomgaard, Peter. "The voc Trade in Forest Products in the Seventeenth Century." In *Nature and the Orient: The Environmental History of South and Southeast Asia* edited by Richard Grove, Vinita Damodaran, and Satpal Sangwan. Delhi: Oxford University Press, 1998.

Bowett, Adam. "The Jamaica Trade: Gillow and the Use of Mahogany in the Eighteenth Century." *Regional Furniture* 12 (1988): 14–57.

Bowett, Adam. "The English Mahogany Trade 1700–1793." PhD dissertation, Brunel University, London, 1996.

Brading, David B. *Bourbon Spain and Its American Empire*. Cambridge: Cambridge University Press, 1981.

Broda, Józef. *Historia leśnictwa w Polsce*. Poznań: Wydawnictwo Akademii Rolniczej im. Augusta Cieszkowskiego, 2000.

Brodacki, Tomasz. "Uwarunkowania prawne ordynacji Radziwiłłowskiej i jej wojsk w Rzeczypospolitej Obojga Narodów." *Zeszyty Naukowe Uniwersytetu Przyrodniczo--Humanistycznego w Siedlcach* 109 (2016): 223–235.

Buczyński, Grzegorz. "Podejście prawne do ochrony lasów w Polsce w ujęciu historycznym." *Kwartalnik Prawa Publicznego* 8 no. 3–4 (2008): 7–36.

Burke, Edmund and Kenneth Pomeranz (eds.). *The Environment and World History*. Berkeley: University of California Press, 2009.

Bustos Rodríguez, Manuel. *Los comerciantes de la Carrera de Indias en el Cádiz del siglo XVIII*. Cádiz: Universidad de Cádiz, 1995.

Bustos Rodríguez, Manuel. *Cádiz en el sistema atlántico: la ciudad, sus comerciantes y la actividad mercantil (1650–1830)*. Madrid: Sílex Ediciones, 2005.

Calderón Quijano, José Antonio. *Historia de las fortificaciones en Nueva España*. Seville: Escuela de Estudios Hispano-Americanos, 1953.

Calderón Quijano, José Antonio (ed.). *Los virreyes de Nueva España en el reinado de Carlos III*. Seville: Escuela Gráfica Salesiana, 1967.

Camille, Michael "The Effects of Timber Haulage Improvements on Mahogany Extraction in Belize: An Historical Geography." *Yearbook (Conference of Latin Americanist Geographers)* vol. 26 (2000): 103–115.

Campillo y Cossío, José del. *Nuevo sistema de gobierno económico para la América: con los males y daños que le causa el que hoy tiene, de los que participa copiosamente España; y remedios universales para que la primera tenga considerables ventajas, y la segunda mayores intereses*. Madrid: Imprenta Benito Cano, 1789 [written in 1743].

Capel Sáez, Horacio. *Los ingenieros militares en España. Siglo XVIII: repertorio biográfico e inventario de su labor científica y espacial*. Barcelona: Universitat Barcelona, 1983.

Carrasco González, Guadalupe. *Comerciantes y casas de negocios en Cádiz, 1650–1700*. Cádiz: UCA, 1997.

Carrasco González, Guadalupe. "Cádiz y el Báltico. Casas comerciales suecas en Cádiz (1780–1800)." In *Comercio y Navegación entre España y Suecia (siglos X–XX)* edited by Alberto Ramos Santana, 317–345. Cádiz: Universidad de Cádiz, 2000.

Casals Costa, Vicente. "Conocimiento científico, innovación técnica y fomento de los montes durante el siglo XVIII." In *El Siglo de las luces: de la industria al ámbito agroforestal* edited by Manuel Silva Suárez, 453–500. Zaragoza: Institución "Fernando el Católico"—Universidad de Zaragoza-Real Academia de Ingeniería, 2005.

Castanedo Galán, Juan Miguel. "Un asiento singular de Juan Fernández de Isla. La fábrica de ocho navíos y la reforma de un astillero." In *El derecho y el mar en la España moderna* edited by Carlos Martínez Shaw, 457–476. Granada: Universidad de Granada, 1995.

Cayuela Fernández, José Gregorio. "Marina española, estrategia conjunta y relaciones internacionales, 1713–1810." In *La Real Armada y El Mundo Hispánico en el Siglo XVIII* edited by Agustín Guimerá and Olivier Chaline, 25–48. Madrid: UNED, 2022.

Chávez, Thomas. *Spain and the Independence of the United States: An Intrinsic Gift.* Albuquerque: University of New Mexico Press, 2002.

Chróściak, Emil. "Szczecin's Maritime Timber Trade and Deliveries to Spain Between 1750 and 1760 on the Basis of Wochentlich-Stettinische Frag- und Anzeigungs-Nachrichten." *Studia Maritima* 33 (2020): 150–164.

Cieślak, Edmund. "Gdański projekt kasy wykupu marynarzy z rąk piratów z połowy XVIII wieku." *Przegląd Historyczny* 51 (1960): 33–51.

Cieślak, Edmund. "The influence of the first partition of Poland on the overseas trade of Gdańska." In *From Dunkirk to Dantzig: Shipping and Trade in the North Sea and the Baltic, 1350–1850* edited by W. G. Heeres et al., 203–215. Hilversum: Verloren, 1988.

Cieślak, Edmund. *Francuska placówka konsularna w Gdańsku w XVIII wieku. Status prawny, zadania, działalność.* Cracow: Polska Akademia Umiejętności, 1999.

Clayton, Lawrence A. *Los astilleros de Guayaquil colonial.* Guayaquil: Archivo Histórico del Guayas, 1978.

Corbera Millán, Manuel. "El impacto de las ferrerías en los espacios forestales (Cantabria, 1750–1860)." *Ería* 45 (1998): 89–102.

Crespo Solana, Ana. *La Casa de Contratación y la Intendencia General de la Marina en Cádiz (1717–1730).* Cádiz: Universidad de Cádiz, 1996.

Crespo Solana, Ana. *El Comercio marítimo entre Cádiz y Ámsterdam, 1713–1778.* Madrid: Banco de España, 2000.

Crespo Solana, Ana. E*ntre Cádiz y los Países Bajos: una comunidad mercantil en la ciudad de la Ilustración, Cádiz.* Seville: Fundación Municipal de Cultura, Cátedra Adolfo de Castro, 2001.

Crespo Solana, Ana. "El patronato de la nación flamenca gaditana en los siglos XVII y XVIII: trasfondo social y económico de una institución piadosa." *Studia Histórica, Historia Moderna* 24 (2002): 297–329.

Crespo Solana, Ana. *Comunidades transnacionales. Colonias de mercaderes extranjeros en el Mundo Atlántico (1500–1830)*. Madrid: Doce Calles, 2010.

Crespo Solana, Ana. "Cádiz y el comercio de las Indias: un paradigma del transnacionalismo económico y social (siglos XVI–XVIII)." *e-Spania* 25 (2016): 1–28.

Crosby, Alfred W. *Ecological Imperialism: The Biological Expansion of Europe, 900–1900*. Cambridge: Cambridge University Press, 1986.

Davey, James. *The Transformation of British Naval Strategy: Seapower and Supply in Northern Europe, 1808–1812*. Suffolk: Boydell Press, 2012.

Davis, Ralph. *The Rise of the Shipping Industry in the Seventeenth and Eighteenth Centuries*. London: Macmillan, 1962.

Dean, Warren. *With Broadax and Firebrand: The Destruction of the Brazilian Atlantic Forest*: Berkeley, University of California Press, 1995.

Demski, Dagnosław. "Naliboki i Puszcza Nalibocka: zarys dziejów i problematyki." *Etnografia Polska* 38, no. 1–2 (1994): 51–78.

Díaz Ordóñez, Manuel. "El riesgo de contratar con el enemigo. Suministros ingleses para la Armada Real española en el siglo XVIII." *Revista de Historia Naval* 80 (2003): 65–74.

Dorta, Enrique M. *Cartagena de Indias: puerto y plaza fuerte*. Bogotá: Fondo Cultural Cafetero, 1988.

Drozdowski, Marian. *Podstawy finansowe działalności państwowej w Polsce 1764–1793: działalność budżetowa Sejmu Rzeczypospolitej w czasach panowania Stanisława Augusta Poniatowskiego*. Warsaw: PWN, 1975.

Dubas-Urwanowicz, Ewa and Jerzy Urwanowicz. *Magnateria Rzeczypospolitej w XVI–XVIII wieku*. Białystok: Wydawnictwo Uniwersytetu w Białymstoku, 2003.

Dubas-Urwanowicz, Ewa and Jerzy Urwanowicza (eds.). *Wobec króla i Rzeczpospolitej. Magnateria w XVI–XVIII wieku*. Cracow: Avalon, 2012.

Einchorn, Karl Friedrich. *Stosunek xiążęgo domu Radziwiłłow do domów xiążęcych w Niemczech uważany ze stanowiska historycznego i pod względem praw niemieckich poltycznych i xiążęcych*. Warsaw: Księgarnia Aug. Emm. Glücksberga, 1843.

Fernández Cañas-Baliña, Ángela. "Tráfico comercial con el Báltico y el Mar del Norte: suecos y daneses, su relación con Cádiz a través del Diario Marítimo de la Vigía, y el Registro de Sund (1789–1800)." *Baetica* 41 (2021): 231–266.

Fernández Duro, Cesáreo. *Armada Española, desde la unión de los reinos de Castilla y Aragón* vol. VIII. Madrid: Sucesores de Rivadeneyra, 1895–1903.

Fernow, Bernhard E. *A Brief History of Forestry in Europe, the United States and Other Countries*. Toronto: Toronto University Press, 1911.

Fraser, David. *Frederick the Great: King of Prussia*. London: Allen Lane, 2000.

Funes Monzote, Reinaldo. "Los conflictos por el acceso a la madera en La Habana: hacendados vs Marina." In *Diez nuevas miradas de historia de Cuba* edited by José Piqueras, 67–90. Castellón: Publicacions de la Universitat Jaume l, 1998.

Funes Monzote, Reinaldo. *From Rainforest to Cane Field in Cuba: An Environmental History since 1492*. Chapel Hill: University of North Carolina Press, 2008.

García Aguado, Juan. *José Romero Fernández de Landa: un ingeniero de marina en el siglo XVIII*. La Coruna: Universidade da Coruña, 1998.

García-Baquero González, Antonio. *Cádiz y el Atlántico (1717–1778): el comercio colonial español bajo el monopolio gaditano*. Seville: Escuela de Estudios Hispano-Americanos, 1976.

García Pereda, Ignacio and Inés González Doncel and Luis Gil Sánchez "La primera Dirección General de Montes (1833–1842).", *Quaderns d'Història de l'Enginyeria* 13 (2012): 209–253.

García-Torralba Pérez, Enrique. *Las fragatas a vela de la Armada Española 1600–1850. Su evolución técnica*. Madrid: 2011.

Gaziński, Radosław. *Handel morski Szczecina w latach 1720–1805*. Szczecin: Wydawnictwo Naukowe Uniwersytetu Szczecińskiego, 2000.

Gaztambide-Geigel, Antonio. "La invención del Caribe en el siglo XX." *Revista Mexicana del Caribe* 1 (1996): 74–96.

Gierszewski, Stanisław *Statystyka żeglugi Gdańska w latach 1670–1815*. Warsaw: PAN, 1963.

Gierszewski, Stanisław. *Wisła w dziejach Polski*. Gdańsk: Wydawnictwo Morskie, 1982.

Glete, Jan. *Navies and Nations: Warships, Navies and State Building in Europe and America, 1500–1860*, vol. II. Stockholm: Almqvist & Wiksell International, 1993.

Glete, Jan. *Warfare at Sea, 1500–1650: Maritime Conflicts and the Transformation of Europe*. London: Routledge, 1999.

González-Aller Hierro, José Ignacio. *Modelos de arsenal del Museo Naval evolución de la construcción naval española, siglos XVII–XVIII*. Barcelona: Lunwerg, 2004.

Grafe, Regina. "The Strange Tale of the Decline of Spanish Shipping" in *Shipping and Economic Growth, 1350–1850*, edited by Richard W. Unger, 81–116. Leiden: Brill, 2011.

Grafenstein Gareis, Johanna von. *Nueva España en el Circuncaribe, 1779–1808. Revolución, competencia imperial y vínculos intercoloniales*. México: Universidad Nacional Autónoma de México, 1997.

Graham, Hamish. "For the Needs of the Royal Navy: State Interventions in the Communal Woodlands of the Lands during the Eighteenth Century." *Journal of the Western Society for French History* 35 (2007): 135–148.

Graham, Hamish. "Fleurs-de-lis in the Forest: 'Absolute' Monarchy and Attempts at Resource Management in Eighteenth-Century France." *French History* 23, no. 3 (2009): 311–335.

Groth, Andrzej. *Rozwój floty i żeglugi gdańskiej 1660–1700*. Gdańsk: Zakład Narodowy Ossolińkich, 1974.

Groth, Andrzej. *Żegluga i handel morski Kłajpedy w latach 1664–1722*. Gdańsk: Wydawnictwo Uniwersytetu Gdańskiego, 1996.

Guimerá Ravina, Agustín. "La casa Milans una empresa catalana en Rusia (1773–1779)." *Pedralbes: Revista d'Historia Moderna* 18, no. 1 (1998): 83–92.

Guldon, Zenon and Lech Stępkowski, *Z dziejów handlu Rzeczypospolitej w XVI–XVIII wieku: studia i materiały,* Kielce: Wyższa Szkoła Pedagogiczna im. Jana Kochanowskiego, 1980.

Harding, Richard. *Seapower and Naval Warfare, 1650–1830.* London: Routledge, 1999.

Haring, Clarence H. *Trade and Navigation between Spain and the Indies in the Time of the Hapsburgs.* Cambridge: Harvard University Press, 2014 [1st edition 1918].

Hausberger, Bernd. *Historia mínima de la globalización temprana.* México: Colegio de Mexico, 2018.

Hedemann, Otton. *Dawne puszcze i wody.* Wilno: Księgarnia Św. Wojciecha, 1934.

Hedemann, Otton. "Wrąb Radziwiłłowski." *Echa leśne* 38 (1935): 4.

Hedemann, Otton. *Dzieje puszczy Białowieskiej w Polsce przedrozbiorowej w okresie do 1798 roku.* Warsaw: Instytut Badań Lasów Państwowych, 1939.

Heeres, W. G., L. M. Hesp, L. Noordegraaf, and R. C. W. Van Der Voort. *From Dunkirk to Dantzig: Shipping and Trade in the North Sea and the Baltic, 1350–1850.* Hilversum: Verloren, 1988.

Hernández Franco, Juan. "Relaciones entre Cabarrús y Floridablanca durante la etapa de aquél como director del Banco Nacional de San Carlos (1782–1790)." *Cuadernos de historia moderna y contemporánea* 6 (1985): 81–92.

Hoffman, Paul E. *The Spanish Crown and the Defense of the Caribbean, 1535–1585.* Baton Rouge: Louisiana State University Press, 1980.

Hutchison, Ragnhild. "The Norwegian and Baltic Timber Trade to Britain 1780–1835 and Its Interconnections." *Scandinavian Journal of History* 37, no. 5 (2012): 578–596.

Inglis, G. Douglas. "The Spanish Naval Shipyard at Havana in the Eighteenth Century." In *New Aspects of Naval History,* 47–58. Baltimore: Department of History US Naval Academy, 1985.

Janzen, Olaf Uwe. *Merchant Organization and Maritime Trade in the North Atlantic: 1660–1815.* Liverpool: Liverpool University Press, 1998.

Jenkins, Ernest H. *A History of the French Navy, from Its Beginnings to the Present Day.* London: Macdonald & Jane's, 1st UK edition, 1973.

Jezierski, Andrzej and Cecylia Leszczyńska. *Historia gospodarcza Polski.* Warsaw: Wydawnictwo Key Text, 2010.

Jiménez-Montes, Germán. *A Dissimulated Trade. Northern European Timber Merchants in Seville (1574–1598).* Leiden: Brill, 2022.

Jofré González, José. "The Sugar Industry, the Forests and the Cuban Energy Transition, from the Eighteenth Century to the Mid-Twentieth Century." In *The Economies of Latin America* edited by César Yáñez and Albert Carreras, 131–146. Cambridge: Cambridge University Press, 2014.

Johansen, Hans. "Scandinavian Shipping in the Late Eighteenth Century in a European Perspective." *Economic History Review* 45, no. 3 (1992).

Jordán Reyes, Miguel. "La deforestación de la Isla de Cuba durante la dominación española: (1492–1898)." PhD dissertation, Universidad Politécnica de Madrid, 2006.

Kamen, Henry. *Spain's Road to Empire: The Making of a World Power, 1492–1763.* London: Allen Lane, 2002.

Kaps, Klemens. "Trade Connections between Eastern European Regions and the Spanish Atlantic during the Eighteenth Century." In *Transregional Connections in the History of East-Central Europe* edited by Katja Castryck-Naumann, 217–258. Berlin: De Gruyter, 2021.

Kargul, Michał. "Administracja leśna w dobrach królewskich w świetle lustracji województwa pomorskiego z 1765 roku." *Acta Cassubiana* 10 (2008): 57–79.

Karonen, Petri. "Coping with Peace after a Debacle: The Crisis of the Transition to Peace in Sweden after the Great Northern War (1700–1721)." *Scandinavian Journal of History* 33 (2008): 203–225.

Kazusek, Szymon. *Spław wiślany w drugiej połowie XVIII wieku (do 1772 roku)* vol. 2. Kielce: Wydawnictwo Uniwersytetu Jana Kochanowskiego, 2016.

Kent, H. S. K. "The Anglo-Norwegian Timber Trade in the Eighteenth Century." *Economic History Review* 8, no. 1 (1955): 64–71.

Knight, Roger J. B. "New England Forests and British Seapower: Albion Revised." *American Neptune* 46 (1986): 221–229.

Knitter, Michał. "Verifizierung von Schifffahrtsstatistiken des Stettiner Hafens in der zweiten Hälfte des 18. und Aufgang das 19. Jahrhunderts." *Studia Maritima* 25 (2012): 23–51.

Kościałkowski, Stanisław. *Antoni Tyzenhauz, podskarbi nadworny litewski*, vols. 1–2. London: Wydawnictwo Społeczności Akademickiej Uniwersytetu Stefana Batorego, 1970.

Kowalski, Mariusz. *Księstwa Rzeczpospolitej. Państwo magnackie jako region polityczny.* Warsaw: IGiPZ-PAN, 2013.

Kuethe, Allan J. and Kenneth J. Andrien. *The Spanish Atlantic World in the Eighteenth Century. War and the Bourbon Reforms, 1713–1796.* Cambridge: Cambridge University Press, 2014.

Kuethe, Allan J. and José Manuel Serrano Álvarez. "El astillero de la Habana y Trafalgar." *Revista de Indias* 67, no. 241 (2007): 763–776.

Kus Józef. "Z dziejów handlu Kazimierza Dolnego w XVII–XVIII wieku: instruktarze cła wodnego z 1616 i 1763 roku." *Rocznik Lubelski* no. 31–32 (1989–1990): 235–241.

Kwapień, Maria and Józef Maroszek and Andrzej Wyrobisz (eds.). *Studia nad produkcją rzemieślniczą w Polsce (XIV–XVIII w.).* Wrocław: Zakład Narodowy im. Ossolińskich, 1976.

Lafuente, Antonio and José Luis Peset. "Política científica y espionaje industrial en los viajes de Jorge Juan y Antonio de Ulloa." *Melanges de la Casa de Velázquez* 17 (1981): 233–262.

Lamikiz, Xabier. *Trade and Trust in the Eighteenth-Century Atlantic World. Spanish Merchants and Their Overseas Networks.* Suffolk: Boydell Press, 2013.

Langton, John and Graham Jones (eds.) *Forests and Chases of England and Wales c.1500–1850: Towards a Survey and Analysis.* Oxford: St. John's College, 2005.

Laviana, María. *Guayaquil en el siglo XVIII. Recursos naturales y desarrollo económico.* Seville: Escuela de Estudios Hispano-Americanos, 1987.

Lesiak, Anna. "Kobiety z rodu Radziwiłłów w świetle inwentarzy i testamentów (XVI–XVIII w.)." In *Administracja i życie codzienne w dobrach Radziwiłłów XVI–XVIII wieku* edited by Urszula Augustyniak, 113–194. Warsaw: Wydawnictwo DiG, 2009.

Lesiński, Henryk. "Przemiany w stosunkach handlowych miast Pomorza Zachodniego w drugiej połowie XVII i początkach XVIII wieku." In *Historia Pomorza (do roku 1815)* vol. 2 edited by Gerard Labuda, 173–216. Poznań: Wydawnictwo Poznańskie, 1984.

Lesiński, Henryk. "Handel morski Szczecina w okresie szwedzkim 1639–1713." *Materiały Zachodniopomorskie* 31 (1985): 277–295.

Lesiński, Henryk. "Rozwój handlu morskiego Szczecina XVI–XVIII." In *Estuarium Odry i Zatoka Pomorska w rozwoju społeczno-gospodarczym Polski* edited by Hubert Bronk. Szczecin: Uniwersytet Szczeciński, 1990.

Lomas Salmonte, Francisco Javier and Rafael Sánchez Saus and Manuel Bustos and José Luis Millán Chivite. *Historia de Cádiz: entre la leyenda y el olvido.* Madrid: Editorial Sílex, 2005.

Lynch, John. *Bourbon Spain, 1700–1808.* Oxford: Basil Blackwell, 1989.

Lynch, John. *Historia de España. Edad moderna: crisis y recuperación 1598–1808,* Volumen V. Barcelona: Crítica, 2005.

Lynch, John. *La España del siglo XVIII.* Barcelona: Crítica, 1991.

Machuca, Laura. "Proyectos oficiales y modos locales de utilización del Istmo de Tehuantepec en la época colonial: historias de desencuentros." In *El istmo mexicano: una región inasequible. Estado, poderes locales y dinámicas espaciales (siglos XVI–XXI)* edited by Emilia Velázquez and Éric Léonard and Odile Hoffmann and M.-F. Prévôt-Schapira, 68–94. Marseille: IRD Éditions, 2009.

McNeill, John R. "Woods and Warfare in World History", *Environmental History* 9, no. 3 (2004): 388–410.

McNeill, John R. and Erin S. Mauldin, *A Companion to Global Environmental History.* Oxford: Wiley-Blackwell, 2012.

Mączak, Antoni (ed.). *Encyklopedia historii gospodarczej Polski do 1945 roku,* tom I. Warsaw: Wiedza Powszechna, 1981.

Mączak, Antoni. *Klientela. Nieformalne systemy władzy w Polsce i Europie XVI–XVIII w.* Warsaw: Państwowy Instytut Wydawniczy, 1994.

Mahan, Alfred T. *The Influence of Sea Power upon History, 1660–1783*. Cambridge: Cambridge University Press, 2010 [1st edition 1889].

Małecki, Jan. *Związki handlowe miast polskich z Gdańskiem w XVI i pierwszej połowie XVII wieku*. Wrocław: Zakład Narodowy Ossolińskich, 1968.

Manyś, Bernadetta. "Przyczynek do badań nad rolą i funkcją lasów w XVIII-wiecznych dobrach radziwiłłowskich w Wielkim Księstwie Litewskim." *Studia i Materiały Ośrodka Kultury Leśnej* 15 (2016): 159–177.

Marchena Fernández, Juan and Justo Cuño (eds.). *Vientos de guerra apogeo y crisis de la Real Armada, 1750–1823*, vol. 2. Madrid: Ediciones Doce Calles, 2018.

Marichal, Carlos. *La bancarrota del virreinato. Nueva España y las finanzas del Imperio español, 1780–1810*. México: FCE, 1999.

Martínez González, Alfredo José. "La elaboración de la Ordenanza de Montes de Marina, de 31 de enero de 1748, base de la política oceánica de la monarquía española durante el siglo XVIII." *Anuario de Estudios Americanos* 71, no. 2 (2014): 571–602.

Martínez González, Alfredo José. *Las Superintendencias de Montes y Plantíos (1574–1748). Derecho y política forestal para las armadas en la Edad Moderna*. Valencia: Tirant Lo Blanch, 2015.

Martínez Shaw, Carlos. "La historia marítima de los tiempos modernos. Una historia total del mar y sus orillas." *Drassana* 22 (2014): 36–64.

Martínez Shaw, Carlos and Marina Alfonso Mola. "Los astilleros de la América colonial." In *Historia general de América Latina* edited by Alfredo Castillero Calvo and Allan J. Kuethe, 279–304. vol. 3, tomo 1. Madrid: Editorial Trotta; Paris: Ediciones Unesco, 2001.

Martínez-Fortún y Foyo, José and Humberto Arnáez y Rodríguez. *Diccionario biográfico remediano*. La Habana: Ayuntamiento de San Juan de los Remedios, 1960.

Matteson, Keiko. *Forests in Revolutionary France: Conservation, Community, and Conflict, 1669–1848*. Cambridge: Cambridge University Press, 2015.

Merino Navarro, José Patricio. *La Armada Española en el siglo XVIII*. Madrid: Fundación Universitaria Española, 1981.

Miller, Shawn W. *Fruitless Trees: Portuguese Conservation and Brazil's Colonial Timber*. Stanford: Stanford University Press, 2000.

Mintz, Sidney. *Caribbean Transformations*. Chicago: Aldine, 1974.

Moncada Maya, Omar. *Ingenieros militares en Nueva España. Inventario de su labor científica espacial. Siglos XVI a XVIII*. México: Universidad Nacional Autónoma de México, 1993.

Moreno Gullón, Amparo. "La Matrícula de Mar de Campeche (1777–1811)." *Espacio, Tiempo y Forma, serie IV, Historia Moderna* 17 (2004): 273–291.

Mosley, Stephen. *The Environment in World History*. Oxford: Routledge, 2010.

North, Michael. "The Export of Timber and Timber by Products from the Baltic Region to Western Europe, 1575–1775." In *From the North Sea to the Baltic: Essays in Commercial, Monetary and Agrarian History, 1500–1800* edited by Michael North, 1–14. Aldershot: Ashgate, 1996.

Nowak, Dorota. "Konsulat hiszpański w Gdańsku w drugiej połowie XVIII wieku. Kilka uwag o możliwościach badawczych." *Roczniki Humanistyczne* 58 (2010): 123–130.

Nyrek, Aleksander. "Stan i praktyki wiedzy leśnej na Śląsku do połowy XIX wieku." *Śląski Kwartalnik Historyczny Sobótka* 3 (1972): 413–432.

Nyrek, Aleksander. *Gospodarka leśna na Górnym Śląsku od połowy XVII do połowy XIX wieku.* Wrocław: Zakład Narodowy im. Ossolińskich, 1975.

Ormrod, David. "Institutions and the Environment: Shipping Movements in the North Sea and Baltic Zone, 1650–1800." In *Shipping Efficiency and Economic Growth, 1350–1800* edited by Richard W. Unger, 135–166. Leiden: Brill, 2011.

Ormrod, David. *The Rise of Commercial Empires. England and the Netherlands in the Age of Mercantilism, 1650–1770.* Cambridge: Cambridge University Press, 2003.

Ozanam, Didier. *Les diplomates espagnols du XVIIIe siècle. Introduction et répertoire biographique (1700–1808).* Madrid: Casa de Velázquez, Maison des Pays Ibériques, 1998.

Palma, Rafael and Odile Hoffmann "La conformación de una frontera interna en las riberas del Tesechoacán." In *Historias de hombres y tierras. Una lectura sobre la conformación territorial del municipio de Playa Vicente, Veracruz* edited by María Teresa Rodríguez and Bernard Tallet, 35–72. México: Centro de Investigaciones y Estudios Superiores en Antropología Social, 2009.

Parry, John H. *The Spanish Seaborne Empire.* Berkeley: University of California Press, 1990.

Parry, John H. and Philip M. Sherlock. *A Short History of the West Indies.* London/New York: Macmillan/St. Martin's Press, 1956.

Pearce, Adrian J. *The Origins of Bourbon Reform in Spanish America 1700–1763.* New York: Palgrave Macmillan, 2014.

Pearson, Charles W. *England's Timber Trade in the Last of the 17th and First of the 18th Century, More Especially with the Baltic Sea.* Whitefish: Kessinger Publishing, 2009 [1st edition 1869].

Perlin, John. *A Forest Journey: The Story of Wood and Civilization.* Woodstock-Vermont: Countryman Press, 2005.

Pezzi Cristóbal, Pilar. "Proteger para producir. La política forestal de los Borbones españoles." *Baética: Estudios de Historia Moderna y Contemporánea* 23 (2001): 583–596.

Pinzón Ríos, Guadalupe. *Hombres de mar en las costas novohispanas. Trabajos, trabajadores y vida portuaria en el Departamento Marítimo de San Blas (siglo XVIII).* México: UNAM, 2018.

Piqueras, José. "El azúcar en Cuba y las fuentes para su estudio." *América Latina en la Historia Económica* 6, no 11 (1999): 35–47.

Plá y Rave, Eugenio. *Tratado de maderas de construcción civil y naval.* Madrid: Imprenta, Estereotipia y Galvanoplastia de Aribau, 1880.

Poschman, Adolf. "El consulado español en Danzig desde 1752 hasta 1773", *Revista de Archivos, Bibliotecas y Museos* 4–6 (1919): 1–24.

Pourchasse, Pierrick. "The control of maritime traffic and exported products in the Baltic area in early modern times: Eighteenth-century Riga." *Revue Historique* 686, no. 2 (2018): 377–398.

Pradells Nadal, Jesús. "Los cónsules españoles del siglo XVIII. Caracteres profesionales y vida cotidiana." *Revista de Historia Moderna* 10 (1991): 209–260.

Radell, David R. and James J. Parsons, "El Realejo: A Forgotten Colonial Port and Shipbuilding Center in Nicaragua." *Hispanic American Historical Review* 51 (1971): 295–312.

Ramos Catalina y de Bardaxí, María Luisa. "Expediciones Científicas a California en el siglo XVIII." *Anuario de Estudios Americanos* 13 (1956): 217–310.

Ramos Santana, Alberto. *Comercio y Navegación entre España y Suecia (siglos X–XX).* Cádiz: Universidad de Cádiz, 2000.

Recio Morales, Óscar, "Los militares de la Ilustración y la construcción del Este de Europa en España." *Itinerarios. Revista de Estudios Lingüísticos, Literarios, Históricos y Antropológicos* 31 (2020): 34–56.

Record, Samuel J. and Robert W. Hess. *Timbers of the New World.* Nueva York: Arno 1972.

Reichert, Rafal. "El comercio directo de maderas para la construcción naval española y de otros bienes provenientes de la región del Báltico sur, 1700–1783." *Hispania. Revista Española de Historia* 76, no. 252 (2016): 129–157.

Reichert, Rafal. "¿Cómo España trató de recuperar su poderío naval? Un acercamiento a las estrategias de la marina real sobre los suministros de materias primas forestales provenientes del Báltico y Nueva España (1754–1795)." *Espacio, tiempo y forma. Serie IV, Historia moderna,* 32, (2019): 73–102.

Reichert, Rafal. "Recursos forestales, proyectos de extracción y asientos de maderas en la Nueva España durante el siglo XVIII." *Obradoiro de Historia Moderna* 28 (2019): 55–81.

Reichert, Rafal. "Direct Supplies of Timbers from the Southern Baltic Region for the Spanish Naval Departments during the Second Half of the 18th Century." *Studia Maritima* 33 (2020) 129–147.

Reichert, Rafal. "El comercio de maderas del Báltico Sur en las estrategias de suministros de la Marina Real, 1714–1795." In *Redes empresariales y administración estatal: la provisión de materiales estratégicos en el mundo hispánico durante el largo siglo XVIII* edited by Iván Valdez-Bubnov and Sergio Solbes Ferri and Pepijn Brandon, 77–94. México: Universidad Nacional Autónoma de México, 2020.

Reichert, Rafal. "El transporte de maderas para los departamentos navales españoles en la segunda mitad del siglo XVIII." *Studia Historica: Historia Moderna* 43 (2021): 47–70.

Rey Castelao, Ofelia. *Montes y política forestal en la Galicia del Antiguo Régimen.* Santiago de Compostela: Universidad de Santiago de Compostela, 1995.

Reyes García Hurtado, Manuel (ed.). *La Armada española en el siglo XVIII: ciencia, hombres y barcos.* Madrid: Silex Ediciones, 2012.

Riezu Elizalde, Óscar and Rafael Torres Sánchez. "¿En qué consistió el triunfo del Estado Forestal? Contractor State y los asentistas de madera del siglo XVIII." *Studia Historica. Historia Moderna* 43, no. 1 (2021): 195–226.

Rodger, Nicholas A. M. *The Wooden World: An Anatomy of the Georgian Navy.* London: Collins, 1986.

Rodger, Nicholas A. M. *The Command of the Ocean: A Naval History of Britain, 1649–1815* vol. 2. New York: Norton, 2004.

Rodríguez González, Agustín Ramón and Juan Luis Coello. *La fragata en la Armada Española: 500 años de historia.* Madrid: IZAR Construcciones Navales, 2003.

Rodríguez González, Agustín Ramón. *Trafalgar y el conflicto naval anglo-español del siglo XVIII.* Madrid: Actas, 2005.

Ruiz García, Vicente. "La Provincia Marítima de Segura (1733–1836). Poder Naval, Explotación Forestal y Resistencia en la España del Antiguo Régimen." PhD dissertation, Universidad de Murcia, 2018.

Salvador y Monserrat, Vicente. *El marqués de Cruilles, Biografía del Excmo. Sr. Teniente general D. Joaquín Monserrat y Cruilles, Marqués de Cruilles, Virrey de Nueva España de 1760 a 1766.* Valencia: Nicasio Rius, 1880.

Sánchez Baena, Juan José, and Celia Chaín Navarro and Lorena Martínez Solís (eds.). *Estudios de historia naval: actitudes y medios en la Real Armada del siglo XVIII.* Madrid: Ministerio de Defensa, 2011.

Sánchez Baena, Juan José, and Cristina Roda Alcantud. "El arsenal del Mediterráneo. Cartagena (1750–1824)." In *Vientos de guerra. Apogeo y crisis de la Real Armada, 1750–1823,* vol. 3 edited by Juan Marchena Fernández and Justo Cuño Bonito, 117–198. Madrid: Doce Calles, 2018.

Sánchez Carrión, José María. *De constructores a ingenieros de marina. Salto tecnológico y profesional impulsado por Francisco Gautier.* Madrid: Fondo Editorial de Ingeniería Naval, 2013.

Schneider, Elena A. *The Occupation of Havana: War, Trade, and Slavery in the Atlantic World.* Chapel Hill: University of North Carolina Press, 2018.

Seijas y Lobera, Francisco de. *Gobierno militar y político del Reino Imperial de la Nueva España (1702).* Mexico: Universidad Nacional Autónoma de México, 1986.

Serrano Álvarez, José Manuel. "Juan de Acosta y la construcción naval en La Habana (1717–1740)." *Revista de Historia Naval* 93 (2006): 7–32.

Serrano Álvarez, José Manuel. "La revitalización de La Habana en época de Lorenzo Montalvo, 1765–1772." *Revista de Historia Naval* 105 (2009): 71–101.

Serrano Álvarez, José Manuel. "Contratas militares en La Habana durante el siglo XVIII: riqueza local y visión imperial." In *Memoria de la Terceras Jornadas de Historia Económica*, 372–382. México: Asociación Mexicana de Historia Económica, 2015.

Serrano Álvarez, José Manuel. *El astillero de La Habana en el siglo XVIII: historia y construcción naval (1700–1805).* Madrid: Ministerio de Defensa, 2018.

Serrano Álvarez, José Manuel. "Élites y política en el astillero de La Habana durante el siglo XVIII." *Obradoiro de Historia Moderna* 28 (2019): 83–104.

Serrano Mangas, Fernando. *Función y evolución del galeón en la Carrera de Indias.* Madrid: Mapfre, 1992.

Siemens, Alfred H. and Lutz Brinckmann. "El sur de Veracruz a finales del siglo XVIII. Un análisis de la relación de Corral." *Historia Mexicana* 26, no. 2 (1976): 263–324.

Smith, Digby. *The Napoleonic Wars.* London: Greenhill, 1998.

Solano de las Aguas, Sergio Paolo. "El Apostadero de la marina de Cartagena de Indias, sus trabajadores y la crisis política de la Independencia." *Economía & Región* 9, no. 1 (2015): 209–243.

Stańczak, Edward. *Kamera saska za czasów Augusta III.* Warsaw: PWN, 1973.

Stanielewicz, Józef "Zarys rozwoju portu i handlu morskiego Szczecina od XVI do XVIII wieku." In *Pomorze Zachodnie w tysiącleciu* edited by Paweł Bartnik and Kazimierz Kozłowski, 121–128. Szczecin: Wydawnictwo AP, 2000.

Stein, Stanley J. and Barbara H. Stein. *Apogee of Empire Spain and New Spain in the Age of Charles III, 1759–1789.* Baltimore: Johns Hopkins University Press, 2003.

Sulimierski, Filip and Władysław Walewski. *Słownik geograficzny Królestwa Polskiego i innych krajów słowiańskich* vol. VIII. Warsaw: Druk Wieku, 1886.

Syrett, David. *The Royal Navy in European Waters During the American Revolutionary War.* Columbia: University of South Carolina Press, 1998.

Taracha, Cezary. "Jeszcze o gdańskiej misji Pedra Arandy w 1761 roku." *Rocznik Gdański* 56, no. 2 (1996): 17–21.

Taracha, Cezary. "El Marqués de la Ensenada y los servicios secretos españoles en la época de Fernando VI." *Brocar* 25 (2001): 109–122.

Taracha, Cezary. "Algunas consideraciones sobre la cuestión rusa y turca en la política española de la época de Carlos III." *Teka Komisji Historycznej* 9 (2012): 53–75.

Taracha, Cezary. *Spies and Diplomats Spanish Intelligence Service in the Eighteenth Century.* Frankfurt: Peter Lang-Internationaler Verlag der Wissenschaften, 2021.

Thurman, Michael E. *The Naval Department of San Blas: New Spain's Bastion for Alta California and Nootka 1767 to 1798*. Glendale: Arthur Clark Co., 1967.

Torrejón Chaves, Juan. "La madera báltica, Suecia y España (siglo XVIII)." In *Comercio y navegación entre España y Suecia (siglos X–XX)* edited by Alberto Ramos Santana, 163–222. Cádiz: Universidad de Cádiz, 2000.

Torres Ramírez, Bibiano. *La Armada de Barlovento*. Seville: Escuela de Estudios Hispano-Americanos, 1981.

Torres Sánchez, Rafael. "La colonia genovesa en Cartagena durante la Edad Moderna." In *Rapporti Genova-Mediterraneo-Atlantico nell'Età Moderna* edited by Rafaele Belvederi, 553–581. Génova: Universita di Genova, 1990.

Torres Sánchez, Rafael. "Contractor State and Mercantilism. The Spanish-Navy Hemp, Rigging and Sailcloth Supply Policy in the Second Half of the Eighteenth Century." In *The Contractor State and Its Implications, 1659–1815* edited by Richard Harding and Sergio Solbes Ferri, 308–335. Las Palmas: Universidad de Las Palmas de Gran Canaria, 2012.

Torres Sánchez, Rafael. "Administración o asiento. La política estatal de suministros militares en la monarquía española del siglo XVIII." *Studia Historica. Historia Moderna* 35 (2013): 159–199.

Torres Sánchez, Rafael. *Military Entrepreneurs and the Spanish Contractor State in the Eighteenth Century*. Oxford: Oxford University Press, 2016.

Torres Sánchez, Rafael. "El estado fiscal-naval de Carlos III. Los dineros de la Armada en el contexto de las finanzas de la monarquía." In *Vientos de guerra. Apogeo y crisis de la Real Armada, 1750–1823* vol. 1 edited by Juan Marchena Fernández and Justo Cuño Bonito, 329–436. Madrid: Doce Calles, 2018.

Torres Sánchez, Rafael. "Los negocios con la armada. Suministros militares y política mercantilista en el siglo XVIII." In *Redes empresariales y administración estatal: la provisión de materiales estratégicos en el mundo hispánico durante el largo siglo XVIII* edited by Iván Valdez-Bubnov and Sergio Solbes Ferri and Pepijn Brandon, 49–76. México: Universidad Nacional Autónoma de México, 2020.

Torres Sánchez, Rafael. *Historia de un triunfo. La Armada española en el siglo XVIII*. Madrid: Desperta Ferro, 2021.

Torres Sánchez, Rafael. "Mercantilist Ideology versus Administrative Pragmatism: The Supply of Shipbuilding Timber in Eighteenth-Century Spain." *War & Society* 40, no. 1 (2021): 9–24.

Trápaga Monchet, Koldo. "El estudio de los bosques reales de Portugal a través de la legislación forestal en las dinastías Avis, Habsburgo y Braganza (c.1435–1650)." *Philostrato: revista de historia y arte* 1 (2017): 5–27

Trápaga Monchet, Koldo. "Guerra y deforestación en el reino de Portugal (siglos XVI–XVII)." *Tiempos Modernos* 39, no. 2 (2019): 396–425.

Trápaga Monchet, Koldo and Félix Labrador Arroyo. "Políticas forestales y deforestación en Portugal, 1580–1640: realidad o mito?" *Ler História*, no. 75 (2019): 133–156.

Trápaga Monchet, Koldo and Álvaro Aragón-Ruano and Cristina Joanaz de Melo (eds.),
*Roots of Sustainability in the Iberian Empires. Shipbuilding and Forestry, 14th–19th
Centuries*. New York: Routledge, 2023.

Trzoska, Jerzy. *Żegluga, handel i rzemiosło w Gdańsku w drugiej połowie XVII i XVIII
wieku*. Gdańsk: Uniwersytet Gdański, 1989.

Urteaga, Luis. *La tierra esquilmada. Las ideas sobre la conservación de la naturaleza en
la cultura española del siglo XVIII*. Madrid: Serbal-CSIC, 1987.

Uztáriz, Jerónimo de. *Theorica y practica de comercio, y de marina: en diferentes discursos y calificados exemplares, que con específicas providencias, se procuran adaptar á
la Monarquia española*. Madrid: Imprenta de Antonio Sanz, 1757 [written in 1724].

Valdez-Bubnov, Iván. *Poder naval y modernización del Estado. Política de construcción
naval española (siglos XVI–XVIII)*. México: Universidad Nacional Autónoma de
México, 2011.

Valdez-Bubnov, Iván. "La representación historiográfica de la guerra en el mar en el
largo siglo XVIII: pensamiento táctico y estratégico, navalismo histórico y metod-
ologías de vanguardia en el siglo XXI.", *Cuadernos dieciochistas*, 21, 2020.

Valdez-Bubnov, Iván "La construcción naval española en el Pacífico sur: explotación
laboral, recursos madereros y transferencia industrial entre Nueva España, Filipinas,
India y Camboya (siglos XVI y XVII)." *Studia Historica-Historia moderna* 43, no. 1
(2021): 71–102.

Valle Pavón, Guillermina del. *Donativos, préstamos y privilegios: los mercaderes y mineros de la Ciudad de México durante la guerra anglo-española de 1779–1783*. México:
Instituto Mora, 2016.

Wachowiak, Bogdan. "Wybrane problemy handlu warciańsko-odrzańskiego w latach
1618–1750." *Przegląd Zachodniopomorski* 26 (2011): 49–66.

Walker, Geoffrey J. *Spanish Politics and Imperial Trade, 1700–1789*. London: Macmillan,
1979.

Warde, Paul. *Ecology, Economy and State Formation in Early Modern Germany*.
Cambridge: University of Cambridge, 2006.

Weber, Klaus. "Germany and the Early Modern Atlantic World: Economic Involvement
and Historiography." In *Beyond Exceptionalism. Traces of Slavery and the Slave Trade
in Early Modern Germany, 1650–1850* edited by Rebekka von Mallinckrodt and Josef
Köstlbauer and Sarah Lentz, 27–55. Berlin: De Gruyter, 2021.

Wing, John T. *Roots of Empire. Forests and State Power in Early Modern Spain,
c.1500–1750*. Leiden: Brill, 2015.

Žiemelis, Darius. "The Structure and Scope of the Foreign Trade of the Polish-Lithuanian
Commonwealth in the 16th to 18th Centuries: The Case of the Grand Duchy of
Lithuania." *Lithuanian Historical Studies* 17 (2012): 91–123.

Index